W9-ASF-212

Science in Nineteenth-Century America
A Documentary History

Science in Nineteenth-Century America

A Documentary History

Edited and with an Introduction by

NATHAN REINGOLD

OCTAGON BOOKS

A DIVISION OF FARRAR, STRAUS AND GIROUX

New York 1979

Reprinted 1979

OCTAGON BOOKS
A DIVISION OF FARRAR, STRAUS & GIROUX, INC.
19 Union Square West
New York, N.Y. 10003

Library of Congress Cataloging in Publication Data

Reingold, Nathan, 1927- ed.
Science in nineteenth-century America.

Reprint of the ed. published by Hill and Wang, New York, in series: American century series.
Includes index.
1. Science—History—United States. I. Title.
Q127.U6R4 1979 509'.73 79-17507
ISBN 0-374-96778-4

Manufactured by Braun-Brumfield, Inc.
Ann Arbor, Michigan
Printed in the United States of America

To

RICHARD H. SHRYOCK,

who first interested me in the history of science

Contents

Introduction

The science of the past is entombed in scattered journals and monographs that bear quaintly archaic titles and contain theories no longer tenable and observations irrelevant to contemporary scientific inquiry. But the workers in the science of the past are nonetheless the creators of the science of today. They are in the main forgotten except by specialists in the history of the sciences; a few great names are still connected with great contributions made to the cultural heritage of Western civilization; other names become immortal as they are affixed to a scientific concept or phenomenon. For the most part, however, the scientists of the past endure only in stilted necrologies, devotional biographies, and solemn portraiture.

Honorable exceptions exist, of course, but until the notable increase of research in the history of science in recent decades, accurate, realistic, and detailed information was hard to come by. Anecdotage there was and also accounts of the growth of the contents of sciences in which human beings and their environments appear almost as an afterthought. What was lacking were studies of science as a living tradition, of scientists participating in the tradition as members of joint intellectual ventures, of the interactions of scientific tradition, institutional setting, individual personalities, and social and cultural milieu.

This collection of documents from nineteenth-century America is a contribution to the attempt to discover and to analyze the origins of our contemporary state of science, in this case the roots in the last century of present American eminence in the sciences. With the exception of the Peirce-Leverrier letters and the final selection, a speech by the physicist Henry Augustus Rowland, the documents that follow were not originally intended for the public eye but were private jottings for friends, colleagues, relatives, or personal recollection. Most are here printed for the first time.

Rendering old manuscripts into print is a hazardous affair. One temptation is to strive for an exact transcription of the handwritten original with all its singularities. Another is a complete rendering into modern usage. Between these two extremes is the admirable textual policies of the editors of *The Papers of Thomas Jefferson*, *The Papers of Benjamin Franklin*, and *The Adams Papers*, who have attempted to preserve the form and spirit of the originals "in a manner intelligible to the present-day reader." With minor exceptions I have followed the textual policy of *The Adams Papers*. Most documents appear in full; others are limited to key passages. Where omissions occur, they are noted. Doubtful readings and any inserted material appear in brackets. At the beginning of each document is a brief identification. At the end of a document or a group of documents from a manuscript collection, the location of the originals is noted. A few items are here in translation into English.

Why nineteenth-century America? That century was a great period in the history of science, but the American contributions, while notable in many instances, were not predominant or pre-eminent in terms of the general advance of the sciences. American attitudes toward science, it is often stated or implied, became evident in the past century and still persist. What is good and what is bad in contemporary American science arise from this past heritage. When Americans fashioned an atomic bomb, the "traditional" American ability to utilize knowledge for practical ends was extolled. When we were left standing on the launching pad gaping at Soviet canines whirling above, traditional American neglect of basic science was to blame. In both instances the truth was not so easily packaged. All this does imply the existence of a thesis on the American scientific heritage, a thesis going back at least to de Tocqueville. In overly simple terms, the thesis is that Americans emphasize applied research at the expense of basic research.

Nineteenth-century science in America is a significant subject for study, then, because it may illuminate a major aspect of the American national character or style and successfully explain trends in our history. It is significant because the traditional thesis implies another—that science advances best when wholly or largely divorced from application. And if science does advance best when divorced from applications (an open question), still another issue arises. What is the relationship between theoretical sciences and applied sciences in technology, medicine, and agriculture?

This book does not answer these large questions. Perhaps they are not answerable questions but rather paradoxes around which historian, philosopher, social scientist, and polemicist will endlessly maneuver. Here is evidence of what it was like to be a scientist in nineteenth-century America. Certain major and minor figures, events, and trends appear that were typical of the century or indicative of things to come. No attempt has been made to cover everyone and everything. The selections inevitably reflect my own personal interests and my successes and failures in locating significant manuscripts.

Within these limitations and the practical requirement of compressing a large amount of material into a single volume, the documents reveal what kinds of scientific activity occurred; the relations of Americans with their European colleagues; the development of self-consciousness in the American scientific community; the growth of certain formal scientific organizations; the relations of the scientists to higher education; and the internal squabbles and jockeying for position and recognition in the scientific world. Pure research is the center of attention; applied fields appear here and there in only occasional references and by indirection.

Many individuals and works have contributed facts, insights, and stimulation to me in the preparation of this work—too many to list. I cannot refrain from acknowledging the kindness of Saul Benison of Brandeis University and Whitfield J. Bell, Jr., of the American Philosophical Society. Marvin W. McFarland of the Library of Congress and A. Hunter Dupree of the University of California made useful comments. While I was visiting at Yale University, William Goetzmann sharpened my understanding of certain points and my students Lawrence Badash, David Musto, and Herbert Winnik clarified some doubtful ones. I owe a special debt of gratitude to the patience and co-operation of the staff of the Science and Technology Division of the Library of Congress. At all times my research was greatly facilitated by the efforts of the staffs of the institutions possessing the original manuscripts: the Library of Congress, National Archives, Smithsonian Institution, National Geographic Society, The Johns Hopkins University Library, American Philosophical Society, Princeton University Library, New York Public Library, Osborn Library of the American Museum of Natural History, Peabody Museum of Yale, Yale University Library, Boston Public Library, Gray Herbarium at Harvard, Harvard University Archives, Houghton Library at Harvard, and Huntington Library. Permission to publish the manuscripts is gratefully acknowledged. Dr. Gilbert Grovesnor kindly allowed publication of the Bell-

Michelson correspondence; Miss Harriette Rowland of Baltimore made available two interesting items about her distinguished father.

Traditionally at this point the author "last but not leasts" his poor wife who has suffered through the gestation period of the volume. We shared the pleasures and vexations of deciphering difficult handwritings, as well as the thrills and disappointments of research. Her skepticism quenched several ill-founded enthusiasms of mine, and her industry kept us moving along. And so last but not least, I wish to thank my wife, Ida H. Reingold, but for whose help this book might still remain thickets of notes and ringlets of microfilm.

I am, of course, solely responsible for the authenticity of the texts, accuracy of facts, and probability of interpretations offered.

TWO GRAND OLD MEN

Benjamin Silliman, Sr.

In 1857 a seventy-eight-year-old gentleman in New Haven, only a few years in retirement from his post at Yale College, started work on an autobiography whose title was consonant with the spirit and style of his age: "Origin and Progress in Chemistry, Mineralogy, and Geology in Yale College and in other places, with personal reminiscences." Nine volumes and five years later his manuscript, which was never published, was completed.

During Silliman's lifetime science in America had grown to a vigorous adolescence and, perhaps just as important to him, science at Yale had advanced from a modest to a commanding position in the American scene. Silliman, although not a great discoverer, accomplished much as a teacher, popular lecturer, and publisher of a scientific journal. The excerpts that follow are from the unpublished autobiography and describe his initiation into the life of science and the founding of his journal.

In 1801 Silliman, a twenty-two-year-old law student with little training in science, was considering a teaching post in Georgia when Timothy Dwight, President of Yale College, asked him to accept the professorship of chemistry and natural history at Yale. Training young minds for the ministry was Yale's specialty, but there were plans afoot to expand the school to the university level. Although Silliman was unqualified when asked, Yale's choice was not surprising. The college wanted a person of unimpeachable orthodoxy, and apparently it sincerely believed that any reasonably able person could acquire scientific expertise. The possibility that Silliman might not turn into a creative investigator was not taken into consideration since he was appointed to teach. Silliman was first sent to study at the University of Pennsylvania and then to Europe. Philadelphia, an Episcopalian-Quaker Babylon to the provincial New Englander, was a new world, not only of science but of culture and good living. Here he met Joseph Priestley,[1] the great English chemist, and Robert Hare,[2] the leading American chemist of his generation. Europe too broadened his horizons. Silliman acquired some of the virtues of the new world of scientific inquiry to add to the better qualities of orthodox New England.

Silliman, whose interests ranged enthusiastically over natural history and chemistry, combined many of the virtues of both the amateurs and the emerging professionals of science. He knew more than the former but not enough to keep up far with the latter. But he could talk and write eloquently about the wonders of nature, which were God's wonders, and thousands crowded the halls in which he lectured and purchased his books.

By far the greatest contribution to science in America was a periodical still in existence and known familiarly for most of its career as *Silliman's Journal*. Through most of the past century the *Journal* was the principal outlet for the research of American scientists as well as their major source of information on developments in Europe. For more than a century Silliman and his family published the magazine as a private venture without benefit of a subsidy from Yale.

The formal titles of this periodical are a neat summary of an important trend in science, especially in America. In 1819 the first volume bore the splendid legend, "The American Journal of Science more especially of Mineralogy, Geology, and the Other Branches of Natural History, including also Agriculture and the Ornamental as well as the useful Arts." To sell his *Journal* Silliman offered a little of something for everyone. The title reflected the interests of the cultivated amateur-professional of the day, like Silliman, to whom all knowledge (whether moral, physical, or social) was science and all arts (both useful and ornamental) were logically connected to science. After the first year and until 1879 the magazine was called the *American Journal of Science and Arts*. During this period the excursions into the humanities, fine arts, the moral sciences, and the social sciences were gradually eliminated in favor of pure science and its applications.

Reflecting the growing away of science from a naïve fellowship with the useful arts, the title became *American Journal of Science* in 1880. However a general magazine publishing research in all sciences could not continue to flourish in a period when scientific specialization produced findings interesting and understandable to only a relatively small number. Highly technical periodicals appeared after 1880, and *Silliman's* began declining from its former eminence. As late as 1932 the *Journal* proudly but inaccurately described itself as "The leading scientific journal in America." Even after World War II it still declared that both natural

[1] Joseph Priestley (1733–1804) was the English chemist who is a claimant for the discovery of oxygen. Because of his unorthodox religious and political views Priestley had to flee to America during the French Revolution.

[2] Robert Hare (1781–1858) was the leading American chemist of the early years of the republic. He taught at the University of Pennsylvania and also did work in electromagnetism and meteorology. Noted for his polemical tendencies.

and physical sciences were its domain. In 1963 it simply is "Devoted to the geological sciences and related fields," an acknowledgement of the obvious. The founder and his successors, his son Benjamin Silliman, Jr.; his son-in-law James Dwight Dana; and his grandson, Edward S. Dana, were most at home intellectually in natural history and the magazine always reflected this bias. But the expansion of scientific specialization had converted a live scientific tradition of amateurism and catholicity to an anachronism.

Excerpts from the Autobiography of Benjamin Silliman, Sr.

INTERVIEW WITH PRESIDENT DWIGHT [OF YALE UNIVERSITY]

While I was deliberating upon this important subject,[3] I met President Dwight, one very warm morning in July 1801, under the shade of the grand trees in the street in front of the college buildings—when after usual salutations we lingered and conversation ensued.

He had been a warm personal friend of my deceased father and their residences being but three miles apart on Holland Hill and Greenfield Hill both in Fairfield, an active interest was maintained between them and their families. The President having, ever, and particularly since his accession to the presidency in 1795, taken a parental interest in the welfare of my brother and myself . . . I felt it to be both a privilege and a duty to ask his advice on this occasion.

After stating the case to him, he promptly replied and with his usual decision said—"I advise you not to go to Georgia; I would not voluntarily, unless under the influence of some commanding moral duty, go to live in a country where slavery is established; you must encounter moreover the dangers of the climate, and may die of a fever within two years; I have still other reasons which I will now proceed to state to you."

FIRST OVERTURE OF THE PRESIDENT IN RELATION TO THE PROFESSORSHIP

He then proceeded to say the corporation of the College had several years before at his recommendation, passed a vote or resolution, to establish a professorship of Chemistry and Natural History as soon as the funds would admit of it. . . .

The time, he said, had now arrived, when the College could safely carry the resolution into effect.

[3] i.e., the Georgia teaching position.

He added however, that it was, at present, impossible to find, among us, a man properly qualified to discharge the duties of the office.

He remarked moreover, that a foreigner, with his peculiar habits and prejudices, would not feel and act in unison with us—and that, however able he might be in point of science, he would not understand our college system and might therefore not act in harmony with his colleagues.

He saw no way but to select a young man worthy of confidence and allow him time, opportunity and pecuniary aid, to enable him to acquire the requisite science and skill, and wait for him until he should be prepared to begin. He decidedly preferred one of our own young men, born and raised among us and possessed of our habits and sympathies.

THE QUESTION MADE PERSONAL

The president then did me the honor to propose that I should consent to have my name presented to the Corporation, giving me, at the same time, the assurance of his cordial support, and of his belief, that the appointment would be made.

I was then approaching 22 years of age, still a youth or only entering on early manhood.

I was startled and almost oppressed by the proposal. A profession that of the law in the study of which I was already far advanced, was to be abandoned, and a new profession was to be acquired, preceded by a course of study and of practice too, in a direction, in which, in Connecticut, there was no precedent.

The good president perceived both my surprise and my embarrassment, and with his usual kindness and resource proceeded to remark to this effect.

"I could not propose to you a course of life and of effort which would promise more usefulness or more reputation. The profession of law does not need you—it is already full, and many eminent men adorn our courts of justice. You may also be obliged to cherish a hope long deferred, before success would crown your efforts in that profession, altho' if successful, you may become richer by the law than you can by science."

OTHER REASONS SUSTAINING THE PROPOSAL

"In the profession which I proffer to you, there will be no rival here. The field will be all your own.

The study will be full of interest and gratification, and the presentation which you will be able to make of it to the college classes and the public, will afford much instruction and delight."

"Our country, as regards the physical sciences, is rich in unexplored treasures, and by aiding in their development you will perform an important public service, and connect your name with the rising reputation of our native land."

"Time will be allowed to make any necessary preparation, and when you enter upon your duties you will speak to those to whom the subject will be new; you will advance in the knowledge of your profession more rapidly than your pupils can follow you and will be always ahead of your audience."

Thus encouraged, by remarks so forcibly put, and so kindly suggested, I expressed my warmest and most respectful thanks for the honor and advantages so unexpectedly offered to me; and asked for a few weeks for consideration and for consultation with my nearest friends. We then emerged from under the shades of those noble elms and I retired thoughtful and pensive to my chamber.

IMPRESSIONS AND REFLECTIONS ON THE OCCASION

The confidence reposed in me by President Dwight, and thus tendered in advance, increased my sense of responsibility, in view of a highly important and arduous undertaking.

I felt it however to be a relief, to escape from the practice of the law, which never appeared to me desirable.

There are indeed bright spots in a career at the bar; right may sometime be vindicated against wrong and injured innocence protected; but the temptation would often be strong, especially when backed by wealth, to contend against justice and by the force of talent and address to make the worse appear the better course, and to screen the guilty from punishment, the fraudulent from the payment that is justly due.

If one could always be engaged in a good cause and could be at liberty to follow the promptings of his conscience, without suppression or perversion of truth, or concealment or palliation of wrong, then indeed the practice of law would appear most desirable and honorable and with requisite talent and learning, and the impulses of a generous temperament, a career at the bar might be truly noble; but having been a diligent and attentive listener in the courts of law during my course of study of the profession, I had seen that the *beau ideal sketch* was too often merely a picture of the imagination.

The associations which the practice of the law created are often highly undesirable.

Often the most unworthy part of mankind throngs the courts of justice or are compelled to appear there by the mandate of law; and the practicing lawyer is obliged to consort with the weak and the wicked, as well as with the wise and good. Such were some of the thoughts which occurred to me on the first view of the question of changing professions.

On the other hand, the study of nature appeared very attractive. In her works there is no falsehood, although there are mysteries, to unveil which is a very interesting achievement. Every thing in nature is strait forward and consistent and there are no polluting influences; all the associations with these pursuits are elevated and virtuous and point towards the infinite creator.

My taste also led me in this direction, and I anticipated no sacrifice of feeling in relinquishing the prospect of practice at the bar although I had no reason to regret that I had spent much time in the study of the noble science of the law, founded as it is in strict reason and ethics, sacred to the best interests of mankind. . . .

HABITS AND SOCIAL INTERCOURSE OF OUR BOARDING HOUSE [1802–03][4]

The gentlemen whom I have named [Horace Binney, Charles Chauncey, Elihu Chauncey, Robert Hare, John Wallace and his brother][5] with the friends and visitors that were by them attracted to the house formed a brilliant circle of high conversational powers. They were all educated men of elevated position in society and their manners were in harmony with their training.

Rarely in my progress in life have I met with a circle of gentlemen who surpassed them in courteous manners in bright intelligence, sparkling sallies of wit and pleasantry and cordial greeting both among themselves and with friends and strangers who were occasionally introduced.

Our hostess Mrs. Smith a high spirited and efficient woman was

[4] In Philadelphia.

[5] Except for Hare these men were lawyers or businessmen. Horace Binney (1780–1875) and the two Chaunceys were originally from New England (Charles Chauncey, 1777–1849, and Elihu Chauncey, 1779–1847), from New Haven in fact. John Bradford Wallace (1778–1837) was a court reporter. His brother Joshua Maddox Wallace, Jr., was a lawyer also.

liberal almost to a fault, and furnished her table even luxuriously. Our habits were indeed in other aspects far from those of teetotalers. No person of that description was in our circle. On the contrary, agreeably to the custom which prevailed in the boarding houses of our cities half a century ago, every gentleman furnished himself with a decanter of wine, usually a metallic or other label being attached to the neck and bearing the name of the owner. Healths were drunk, especially if stranger guests were present and a glass or two was not considered excessive, sometimes two or three according to circumstances . . .

I do not remember any water drinker at our table or in the house, for, total abstinence was not then thought of except perhaps by some wise and farseeing Franklin.

EFFECTS OF THIS LUXURIOUS LIVING ON MY HEALTH

Accustomed to a simple diet in New Haven without wine or porter and perhaps with only cider at dinner, the new life to which I was now introduced did not agree well with my health . . . In my case the effects of luxurious living were to a degree counteracted by vigorous exercise. . . .

ABSENCE OF RELIGIOUS OBSERVANCES

There was no outward manifestations of religion in our boarding house.

Grace was I believe never said at table nor did I ever hear a prayer in the house. I trust that private personal prayers ascended from some hearts and lips in a house where so many were amiable and worthy although without a religious garb.

The sabbath. Some of our gentlemen resorted to the churches and some dined out on that day . . .

Friendship and aid of Robert Hare, AFTERWARDS PROFESSOR

The deficiencies of Dr. [James] Woodhouse's[6] courses were, in a considerable degree, made up in a manner which I could not have anticipated.

I have already mentioned that Robert Hare was a fellow boarder and companion at Mrs. Smith's.

He was a genial kind hearted man one year younger than myself and was already a proficient in Chemistry upon the scale of that period and being informed of my object in coming to Philadelphia,

[6] Silliman's chemistry teacher at the University of Pennsylvania.

he kindly entered into my views and extended to me his friendship and assistance.

A SMALL WORKING LABORATORY

The Hydro oxygen blow pipe was conceded to us by the indulgence of our hostess Mrs. Smith and we made use of a spare cellar kitchen in which we worked together in our hours of leisure from other pursuits.

Mr. Hare had, one year before, perfected his beautiful invention of the oxy hydrogen blow-pipe and had presented the instrument to the chemical society of Philadelphia. His mind was much occupied with the subject and he enlisted me into his service.

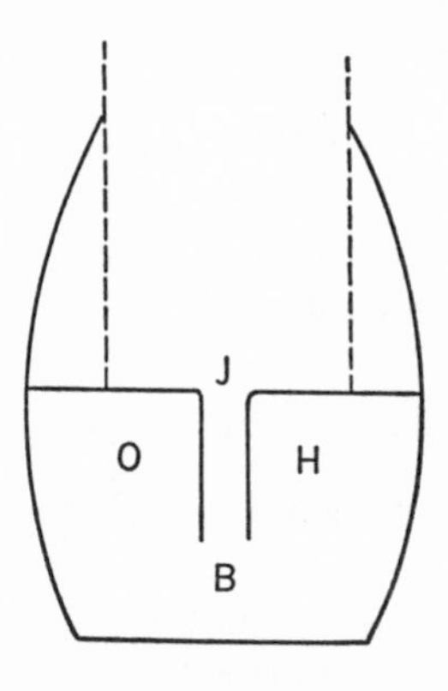

We worked much in making oxygen and hydrogen gases burning them at a common orifice to produce the intense heat of the instrument.

Hare was desirous of making it still more intense by deriving a pure oxygen from chlorate of potassa, then called oxy muriate of potassa [i.e., potassium chlorate]. Chemists were then ignorant of the fact that by mixing a little oxide of manganese with the chlorate the oxygen can be evolved by the heat of a lamp applied to a glass retort.

Hare thought it necessary to use stone retorts with a furnace heat, the retorts were purchased by me at a dollar each and as they were usually broken in the experiment, the research was rather costly but my friend furnished experience and as I was daily acquiring it I was rewarded both for labor and expense by the brilliant results of our experiments.

Hare's apparatus was ingenious but unsafe as regards the storage of the gases. A barrel with a bottom but without a top had a diaphragm across the middle with a metallic tube J. and a vertical partition descending the length of the tube—thus dividing the space into two compartments, the one destined to contain oxygen, *O*—the other hydrogen gas, *H*. Hydrostatic bellows at B by an ingenious arrangement conveyed the gases separately to their respective cells and they were emitted by tubes inserted where the dotted lines are and these were made to terminate in a common orifice. Hydrostatic pressure governed by the quantity of the gases gave them the requisite afflux and they were fired at the point of issue (not here represented).

Novice as I was I ventured to suggest to my more experienced friend that by some accident or blunder the gases, near neighbors' as they were in their continguous apartments, might become mingled when on lighting them at the orifice an explosion would follow. I was afterwards informed although not by Hare that this accident actually happened to him although with no other mischief than a copious shower bath from the expulsion of the water. . . .

MEETING WITH DR. JOSEPH PRIESTLEY

This celebrated gentleman was also a guest on one of the occasions when I dined at Dr. Wistar's.[7] As a very young man (23-24 years) I felt it an honor and advantage to be introduced to so celebrated an author and philosopher.

He had become obnoxious in his native country on account of his political and religious opinions as he was a friend of civil liberty and his religious creed was arian or unitarian.

At that time during the early part of the French revolution there was a stormy excitement in England against revolutionary sentiments and movements. Dr. Priestley then resided at Birmingham and during an anniversary commemoration of the destruction of the Bastille, although he was not then in the city, the mob proceeded to his house which they burned with his library, apparatus and manuscripts: all were lost and the outrage was said to have been countenanced by persons of consideration both lay and clerical. In 1794 he fled from persecution and took refuge with his family at Northumberland Pennsylvania on the Susquehannah river. Here he resumed his philosophical pursuits and made occasional visits to Philadelphia.

[7] Dr. Caspar Wistar (1761–1818) was professor of anatomy, midwifery, and surgery at the University of Pennsylvania. Silliman attended his course.

It was on one of these occasions that I was invited to meet him at Dr. Wistar's table and the interview was to me very gratifying.

In person he was small and slender and in general outline of person not unlike the late President Stiles [of Yale].

His age was then about 70.

His dress was clerical and perfectly plain.

His manners were mild, modest and conciliating so that although in controversy a sturdy combatant he always won kind regard and favor in his personal intercourse. At the dinner Dr. Priestley was of course the honored guest and there was no other except one gentleman and myself.

Some of Dr. Priestley's remarks I remember: speaking of his chemical discoveries which were numerous he said. "When I had made a discovery I did not want to perfect it by a more elaborate research but at once threw it out to the world that I might establish my claim before I was anticipated." He remarked upon those passages in the Epistle John which relate to the Trinity that they are modern interpolations, not being found in the most ancient manuscripts.

He spoke much of Newton and his discoveries and the beauty and simplicity of his character and I think that he claimed him as thinking in religion as he himself did.

He mentioned being present at a dinner in Paris given by the Count de Vergennes during the American revolution and the seat next to him was occupied by a French nobleman.

At another part of the table were two gentlemen dressed in Canonicals, when said Dr. Priestley I enquired of the nobleman the names of those two gentlemen, he replied, "O, one of them is bishop so and so and the other bishop so and so; but they are very clever fellows and although they are bishops they do not believe anything more of this mummery of Christianity than you or I do."

"Speak for yourself sir I replied, for, although I am accounted a heretic in England I do believe what you call this mummery of Christianity". . . .

AMERICAN JOURNAL OF SCIENCE AND ARTS

In following up as far as possible the annals of our scientific labors, we now come to the birth of the work named above. In the preface to the 50th Volume of that work being the Index volume of the entire series to that time, 1846, there is a full history of the rise and progress of the Journal. It is not my design to recapitulate it on this occasion except so far as to mark its origin at this era.

Dr. Archibald Bruce of New York had in 1810 instituted an American Journal of Mineralogy; it was ably conducted and was most favorably received, but it lingered with long intervals between its four numbers and stopped with one volume of 270 pages.

The declining health of Dr. Bruce ending in apoplexy, rendered any prospect of the continuance of his Journal hopeless while his own life being in doubt, and was actually ended on the 22d day of February 1818 in the 41st year of his age. Anticipating the death of Dr. Bruce and it being certain that his Journal could never be revived by him, Col. George Gibbs in an accidental meeting on board the Steamer Fulton on Long Island Sound in 1817 urged upon me the duty of instituting a new Journal of Science that we might not only secure the advantages already gained but make advances still more important.

After much consideration and mature advice I reluctantly consented to make the attempt. It was not done however without shewing due deference to Dr. Bruce. It was in the autumn of 1817 that I called upon him at his house and asked his opinion which was given at once in favor of the effort and moreover in approbation of the plan which included the entire circle of the Physical Sciences and their applications. The first number appeared in July 1818 and the Journal under many discouragements and through some perils has survived until this time February 3, 1859 having already had a life of 40½ years, and the labors of its editors and contributors are recorded in 76 volumes.

Silliman Papers, Yale University Library

Nathaniel Bowditch

If Silliman the popularizer was typically American, so too was his contemporary, Nathaniel Bowditch (1773–1838). Bowditch was a poor boy who rose to a position of responsibility in the business world. But unlike the heroes Horatio Alger later wrote about, Bowditch spent most of his adult life in the grip of a grand passion for mathematics and astronomy.

Bowditch was born in Salem, Massachusetts, then a thriving seaport. In 1785 he went to sea. While on long ocean voyages he studied mathematics and foreign languages. In a few years young Bowditch commanded his own ship, and given his intellectual interests, he quite naturally tried his hand at applications of these fields to navigation. The result was *The New Practical Navigator* (1802). Quickly superseding its predecessors, it became a standard reference work for seamen throughout the world which, in revised edition, is still widely used. During his life Bowditch continued to educate himself; at forty-five he learned German in order to read the important scientific literature then beginning to appear in that language.

After leaving the sea in 1803 Bowditch engaged in business, eventually becoming the administrator of various trusts and estates. He moved from Salem to Boston, where he became an official of insurance companies; the positions he held enabled him to apply his mathematical ability to actuarial problems. Business responsibilities did not detract from his preoccupation with scientific work. Indeed, his success in business enabled Bowditch to complete his greatest scientific contribution.

Astronomy had great prestige in America in Bowditch's day but few practitioners, none matching him in mathematical skill. The best known American astronomer of the eighteenth century, David Rittenhouse[8] of Philadelphia, was already dead when the *Navigator* appeared. In Bowditch's letter of June 11, 1802, to Professor Robert Patterson (1743–1824) of the University of Pennsylvania and an officer of the American Philosophical Society, (which appears below), he unintentionally called attention to some of the differences between him and the older man (who was also largely self-taught). Not only was Bowditch superior as a mathematician, he was genuinely learned in the literature of his field. Eventually this combination of native ability, erudition, and industry would bring Bowditch much prestige in Europe for his contributions.

Although very talented, Bowditch's work was only modestly original. The letter of September 18, 1805, to the French mathematician Sylvestre-François LaCroix[9] (which appears below) provides an example of his principal role in science—that of a very able and learned corrector and extender of the work of others. A masterful example in this vein is Bowditch's translation of the *Mécanique céleste* of the great French mathematician, Pierre Simon, Marquis de Laplace,[10] a work widely acclaimed as the culmination of the Newtonian revolution. The first four volumes of Laplace's work appeared in 1814–17; Bowditch translated them as they came out. He never finished the fifth volume which only was published near the end of his life. Bowditch did more than a literal translation; he corrected errors, gave omitted mathematical proofs, inserted explanatory commentaries, and took pains to show where Laplace was using the work of others without acknowledgment. The last was per-

[8] Rittenhouse (1732–96) was a highly competent astronomer with great skill in devising and constructing instruments. He served as Director of the Mint, a post which became a kind of sinecure for the leading scientists following the precedent of Newton.

[9] LaCroix (1765–1843) was then teaching at the École Polytechnique. He produced many valuable textbooks in mathematics.

[10] Laplace (1749–1827) was one of the great mathematicians and mathematical astronomers.

haps a sufficient reason why Laplace never once replied to Bowditch's letters.

To reach his peers in Europe Bowditch carefully distributed printed copies of the translation. When, in 1830, a friend, Reuben Dimond Mussey,[11] went overseas, he carried volumes for European scientists. In France Mussey relied on an American expatriate, David Bailie Warden,[12] who frequently acted as a link between American and European intellectuals, for proper introductions to the scientific community. As a result Bowditch had the pleasure of receiving a cordial letter from A. M. Legendre (1752–1833), the mathematician he regarded as the greatest of the day. How pleased he must have been to receive at last a letter from a Laplace, the son if not the father. Did he perhaps reflect on the irony of Laplace's advising him, an insurance executive, to work on probability theory?

The favorable response from overseas must have especially gratified Bowditch because he had published the volumes of the *Mécanique céleste* translation at his own expense. No one else in America would publish the work without subsidy, and Bowditch refused to beg or to ask people to subscribe out of consideration for him to a work they could not read. The total cost of printing the four volumes was $12,000 or about one third of his estate at his death in 1838.

What was lacking in the United States then that placed so heavy a burden on a scientist? There was simply no proper scientific milieu, especially for a Bowditch. For many decades American scientists were hampered by many factors in their efforts to create a scientific community. One factor was an irrelevant nationalism; for example, William Lambert,[13] who is mentioned in the Jefferson letter to Bowditch of May 2, 1815, was an ardent advocate of running the prime meridian through Washington, D. C., rather than Greenwich. Bowditch publicly ridiculed the proposal and Lambert's astronomical work. Yet a foolish nationalism often dissipated what little scientific resources there were. A second factor appears in Benjamin Vaughan's[14] letter of August 13, 1817—a pervasive localism which some times hobbled attempts at effective organization and cooperation.

Bowditch himself was greatly handicapped by his solitary intellectual position in his native land, a man without any true predecessors or successors. The American academic world had little to offer him in this

[11] Mussey (1780–1866) was a physician then teaching at Dartmouth.

[12] Warden (1772–1845) was English-born but an American citizen who briefly was consul in Paris.

[13] Not much is known about Lambert besides his espousal of an American prime meridian. Others in America and elsewhere wanted to have a non-British prime meridian.

[14] Benjamin Vaughan (1751–1835) was an Englishman with views like Priestley's. He came to America, settling in Maine. His brother John (1775–1841) served for many years as secretary of the American Philosophical Society, the oldest and leading learned group in America.

respect; they were busy with the instruction of boys. During his lifetime Bowditch turned down offers to teach at Harvard, West Point, and Virginia. (The offer to come to the latter is in Jefferson's letter of October 26, 1818, appearing below.) In Europe there were more men who could appreciate Bowditch and who might have spurred him to greater achievements. There too were a number of scientific institutions aiding investigators, often with funds from governments and the wealthy. Nothing comparable existed in America.

Silliman, starting his career in the same year that Bowditch published *The New Practical Navigator,* made a more practical choice of fields. By inclination and by virtue of a lesser talent, his were not as rigorous and were more easily communicable. More people could flock to study with Silliman, and there was promise of greater immediate practical benefits. A Silliman had successors who could carry forward his favorite sciences. Bowditch was working in a field then seemingly close to perfect completion. The great days of Silliman's sciences were ahead and thus a greater challenge. Nevertheless, a considerable portion of the history of the sciences in the United States in the last century consists of movements to introduce greater rigor into natural history and to overcome the serious gap created by the absence of a significant body of mathematically oriented physical scientists.

Nathaniel Bowditch to Robert Patterson

Salem (Massachusetts) June 11, 1802

Sir.

Though personally unacquainted I have taken the liberty of addressing to you the enclosed demonstration of a theorem given by Mr. Rittenhouse in the 3rd volume of the Transactions of the "American Philosophical Society", for finding the sum of the powers of the sines. Having accidentally met with that paper, I found on examination that it was only a simple corollary of a much more general proposition published by Newton in his tract "De Quadratura Corvarum", and by many other authors; as De Moivre, in the Philosophical Transactions; Simpson in his Fluxions; Harris in his "Lexicon Technicum" etc. some of which were published near a century ago.

Mr. Rittenhouse observes that he had not been able "strictly to demonstrate any more than the first two cases," that is, for the 1st and 2d powers, the four subsequent cases he demonstrated by Infinite Series, and then, by comparing them, concluded the Law of continuation of the higher powers. At the close of the paper, Mr. Rittenhouse requests his corres[pondent] to furnish a demonstration of the 3rd or any subsequent case; but as nothing of the kind has been given in the late memoirs of the Society, I thought it probable that nothing had been presented, and therefore drew up the enclosed paper, which you may present to the Society, if you think proper, in order for it to remain on the files, for it ought not to be published in the memoirs, being only a precise demonstration of a simple case of the general theorem discovered and demonstrated by Newton of the above 100 years ago.

Accept this, Sir, from a young man (an American) who without any education except such as could be procured by himself during a few hours spared from a busy employment hopes that if this trifle does not meet with approbation it will be excused for the consideration of the difficulties the writer labored under in persuing his studies.

With respect, Sir, I remain your humble servant.

Nathaniel Bowditch

P.S. I would thank you to inform me if this come safe to hand.

N. B.

Bowditch to LaCroix

Salem September 18, 1805

Sir.

By Mr. John Cabot who sails soon for Europe I have an opportunity of sending you a few notes on your admirable work sur le calcul integral.

In publishing a work of that kind, arranged in a new form, with many new symbols, it was impossible to avoid typographical errors. Such errors are of small moment and scarcely ever cause any embarassement to the reader. In going over your work, which I read with great attention, profit and pleasure, I discovered several small typographical faults, which though trifling in themselves, I thought might be useful if you were about publishing another edition of your work and from its great merit I had no doubt that would be the case. This was the only inducement I had in forwarding you these papers. You must accept them as the offering of a young man and *one of your scholars* as the homage of his respect for the instructions he had received from your works. You must excuse any faults in these notes, as it is very probable I may have noted for erroneous places which were in fact correct.

The leisure time I had on two or three India voyages I employed myself frequently for determining the longitude by lunar observations. And not withstanding the immense number of formula already given for reducing the apparent to the true distance, yet the problem is of such great importance in navigation that any further simplification is deserving of notice. Whether what I have done in the work lately published will be esteemed as an improvement on the method of approximation, must be left to the public to determine. I have tendered you a copy of the method with its demonstration. The main advantage of my method consists in being *entirely free from cases;* all the corrections being additive, which is not the case in the methods of approximation I have seen and the greatest advantage in the direct methods (such as those of Jean Charles Borda, Richard Dunthirne etc) is the uniformity arising from not being embarassed with a variety of cases, though Borda's method requires 7 places of figures in the seconds which renders the calculations long.

I should esteem it a great favor if you would examine the method if it would not encroach too much on your time.

I would thank you to write to me whether these papers come safe to hand if they will be of any service to you. A letter directed as below and sent to the United States by any vessel sailing from any of the ports of France will if it arrives in this country, come safe to hand.

I am with great respect your most obedient humble servant

Nathl Bowditch

Thomas Jefferson to Bowditch

Monticello May 2, [18]15

I thank you, Sir, for your highly scientific pamphlet on the motion of the pendulum, and more particularly for that containing the deductions of longitudes of places in the United States, from the Solar eclipse of 1811. That of Monticello is especially acceptable, having too long lost familiarity with such operations to have undertaken it my self. Mr. Lambert of Washington had also favored me with his calculation, which varied minutely only from your's: he having, from the same elements, made the Longitude of Monticello 78° 50′ 18″.877 W. from Greenwich. I am happy indeed to find that this most sublime of all sciences is so eminently cultivated by you, and our Rittenhouse was not the only meteor of the hemisphere in which he lived.

Accept the homage due to your science from one who is only a dilettante, and sincere wishes for your health and happiness.

Th. Jefferson

Benjamin Vaughan to Bowditch

August 13, 1817

My dear Sir,

Perhaps I take more pleasure in stating, than you will in hearing, that your certificate is signed, preparatory to your being voted for as an honorary member of the Royal Society of London. Sir Joseph Banks (with whom this is not usual) was the first to sign; and, among others, Dr. Herschel, Mr. Pond the Astronomer Royal, and Sir Charles Blagden,[15] have added their names. The elections take place (according to the forms of the Society) in Easter 1818; when your election is morally certain. No one can do you more justice than Sir Joseph Banks. The above intelligence I have from my brother W[illia]m, who is a member of the Royal Society, and has been witness to the whole proceeding.

Permit me now to revert to the subjects of our last conversation, namely the necessity of giving a *stimulus* to science in Massachusetts, and the propriety of the establishing a *reputation* in foreign countries for attachment to science on the part of the U[nited] States.

1°. Massachusetts ought to be the "headquarters of science" in the U. S., on acc[oun]t of the numbers of its citizens who are *well educated* (exceeding those of any other of our states), as well as to make up for its natural deficiencies. Yet, excepting as regards the progress of Harvard College, Massachusetts makes little or no effort. Its American Academy is asleep perhaps in the sleep of death; and while it sleeps, other petty rivals take (as it were) from between its teeth, the materials which ought to go to its support. Instead of being the organ of New England's learning, its own journals and its own petty societies carry away what ought to belong to it; and Connecticut has instituted a rival Society. So far from our dividing the force of *Massachusetts*, it ought to combine the whole of *New England* in favor of knowledge. Massachusetts will not be allowed to be the *representative* of New England; but it may be the chief mover in a *central society;* other states sending representatives, with

[15] Sir Joseph Banks (1743–1820) was president of the Royal Society and a noted patron of science, especially natural history. Sir William Herschel (1738–1822) was a famous astronomer noted for his improvements in the telescope and for his ambitious plan, surprisingly successful, to survey the entire heavens, as well as many discoveries. John Pond (1767–1836) the Astronomer Royal since 1811 was an eminent practical astronomer. Sir Charles Blagden (1748–1820) was a physician and the secretary of the Royal Society. The names of these men, all of the highest standing, was a clear indication of Bowditch's high reputation.

authority to select pieces for publication belonging to the whole; and philosophical and literary societies of the other states, and (in default of these) their colleges may provide these representatives. But if it should be thought too troublesome to send round a deputation of two or three gentlemen to organize this business, with *all* or a *part* of the New England states; the effort must then be made in Massachusetts and Maine; remembering that no separate *society* and no separate publication in the joint state can singly have magnitude or interest enough to attract general notice in the U. S. and much less in Europe. The *whole* of our philosophers and literary men *united* can only do it by making exertions in unison with each other. The American Academy by the suspension or slow rate of its publications, and by its want of zeal in soliciting pieces; must receive the chief share of censure for what has occurred on this occasion. An annual publication would satisfy the impatience of authors and the public; and thus prevent many contributions from being diverted from this channel, and seeking the public notice through other publications. But where would be the difficulty of the Linnaean Societies having a *department* in this publication alloted to it? The Agricultural Society also, when it has a piece which it conveys to the world natural knowledge, and the Connecticut Society may be provided with similar opportunities. The printing might be so contrived, that each Society might have copies to bind up separately, and make volumes of its own, when its papers admitted of it. Doubtless the adjunct Societies ought to pay their share of the expenses of such a publication; unless the states and individuals could be prevailed upon to assist. But something ought to be done in some way. I will simply add that the Royal Societies of Montpelier and Paris formerly published their memoirs in one and the same Volume; keeping them apart; which is precedent enough for what is here proposed. For myself I perceive no real objection to collecting the published pieces of New England on the subjects in question, yearly; and annexing them (when consent is obtained) to the Massachusetts Volume: but if that be thought degrading let pieces which can not otherwise be obtained, be first presented to the Academy, then published elsewhere with the *leave* of the Academy (affixed to the head of it), and be resumed by the Academy at the end of the year, with the corrections of the author and the remarks of others. The Academy would maintain an easy ascendancy on this occasion; for, if its own contributions did not exceed *all* of the others, they would at least exceed those of *any* of the others; and it would besides have the merit of giving to the whole a currency which others could not expect by their separate efforts. As matters stand arranged, little can be expected from the dislocated action of each state, in consequence of the favors of each branch of knowledge acting separately. All the *petty* journals and philosophical transactions of New England have a local circulation, and possess an ephemeral existence; for nothing can be widely extended and lasting, which is not both well supported, and important, and also of a size at its first publication to be considered as a *book*. But if this must be an insulated effort in favor of science on the part of Massachusetts, at least let the Academy do what the case admits: let it rally its forces, and recruit them. A circular letter calling for contributions from members or their friends; and a vote for doubling the members of the Society (which the state of our *literary* society will well admit); are among the measures to be employed. On this occasion your *imprimis* will be looked to for contributions. I know of no person with a fertile mind who has not embryo ideas on various subjects, which a little exertion will develop; and the transactions of a Philosophical Society may be considered in the first instance as the mere depôt of *fugitive* pieces, which are not yet of size enough to form separate works. But I proceed to my next subject.

2⁰. Reputation is of use to a nation from whatever source offered. France is a single proof of this worth all others; for had it not been for her authors, she never would have obtained the ascendancy she has long possessed in the world. Her authors now stand next to her military men in influence in France itself, and perhaps would go far to govern some of her military men. Bonaparte too was in league with the literature of France, and found in it his best support, next to arms and to bribery. You will not dispute the *uses of literature* to the U. S. as a nation, and I only speak here of the benefit of a *reputation for literature* in foreign countries.

It is with this view then, that I wish you to exhibit some of the errors and deficiences of our friends in Europe on mathematical subjects which have occurred to you; that these persons may at least be made more modest in their carriage towards us. This is more likely to influence their feelings, as well as their judgements, than minor new discoveries originating here. You will do this with a gentleness and propriety which will add to the force and extent of the lesson. And pray, my good sir, when is Europe to begin to think us out of our non-age or rather out of that state of nonentity in which we have so long remained unless by observing some signs of scientific existence in us exhibited in overt-acts? When you have done this, you will find then civility to you and to the nation redoubled. These little incidents also are always matters for historical notice. Besides, your friend Lord Bacon says *Serpens non fit draco nisi serpentem devoraverit* [A serpent does not become a dragon except that it shall have devoured a serpent].

3^{0}. I cannot conclude without mentioning the reproachful disregard we pay to our own authors. No American has published a collection of Dr. Franklin's work; and what has been done on this subject by British editors, is scarcely complete. Till within a few weeks, no one was sure or at least was disposed to make others sure, *where* Dr. Franklin was born in Boston. Again: there is no collection any where made of Count Rumford's works. This should be given as a task to Dr. [Jacob] Bigelow[16] by the Academy, and Harvard College and other societies and colleges; and thus a national effort ought to be made to support the addition. Is it not disgraceful, that nothing should be done towards publishing Count Rumford's works on Caloric, when he has offered a prize on the subject; (a prize indeed not yet contended for nor thought of here)? No one here also knows of [Edward] Bancroft's work on Colors, and few of [William Charles] Wells on Dew: and yet both are Americans. Rittenhouse is neglected also.[17] All these work ought to be printed in quarto, for the sake of the plates: and to be uniform; that those who buy one, may have a motive for buying the whole.

Jefferson to Bowditch

Monticello Oct. 26, [18]18

Dear Sir

I have for some time owed you a letter of thanks for your learned pamphlet on Dr. Stewart's formula for obtaining the sun's distance from the motion of the moon's Apsides; a work however, much above

[16] Jacob Bigelow (1786–1879) was a botanist who also held the Rumford Chair at Harvard. He is credited with inventing the term "technology."

[17] Of the five mentioned, three (Benjamin Thompson, Count Rumford; William Charles Wells; and Edward Bancroft) were Tories who left the United States. Vaughan had edited two early collections of Franklin's writings, the last in 1806 entitled "complete" but not so in Vaughan's opinion in 1817. Wells is better known today for a statement of the theory of natural selection in 1813 than his study of dew. Bancroft served as a spy for both sides during the Revolution. Rumford is significant for his role in the origin of the concept of the mechanical equivalent of heat.

my Mathematical stature. This delay has proceeded from a desire to address you on an interest much nearer home, and on the subject of which I must make a long story.

On a private subscription of about 50 or 60,000 Dollars we began the establishment of what we called Central college, about a mile from the village of Charlottesville and 2 miles from this place, and have made some progress in the buildings. The legislature, at their last session, took up the subject and passed an act for establishing an University, endowing it for the present with an annuity of 15,000 Dollars and directing Commissioners to meet, to recommend a site, a plan of buildings, the professorships necessary for teaching all the branches of science at this day deemed useful etc. The Commissioners, by a vote of 16 for the Central college, 2 for a 2nd place and 3 for a 3rd adopted that for the site of the University. They approved, by a unanimous vote, the plan of building begun at that place, and agreed on such a distribution of the sciences it was thought would bring them all within the competence of 10 professors; and no doubt is entertained of a confirmation by the legislature at their next meeting in December. The plan of building is, not to erect one single magnificent building to contain every body and every thing, but to make of it an Academical village, in which every professor shall have his separate house, containing his lecturing room, with 2, 3 or 4 rooms for his own accommodation, accordingly as he may have a family or no family, with kitchen, garden Etc. Distinct Dormatories for the students, not more than two in a room, and separate boarding houses for dieting them by private housekeepers. We conclude to employ no Professor who is not of the 1st order of the science he professes, that where we can find such in our own country, we shall prefer them, and where we cannot, we will procure them where ever else to be found. The standing salary proposed is of 1000. to 1500. Dollars with 25. Dollars tuition fee from each student attending any professor, with house, garden Etc. free of rent. We believe that our own state will furnish 500 students, and having good information that it will be the resort of all the Western and Southern states, we count on as many more from them, when in full operation. But as the schools will take time to build, we propose until the tuition fees, with the salary shall amount to 2000. Dollars, we will make up that deficiency so as to ensure 2000. Dollars from the out-set. The soil in this part of our country is as fertile as any upland soil in any of the Maritime states, inhabited by a substantial yeomenry of farmers (tobacco long since given up) and being at the 1st ridge of mountains, there is not a healthier or more genial climate in the world. Our maximum of heat, and that only of 1 or 2 days in summer is about 96. The minimum in winter is + 5½ but the mean of the months June July Aug. are 72, 75, 73, and of Dec. Jan. Feb. are 45, 36, 40. The thermometer is below 55. (the fire point) 4. months in the year, and about a month before and after that we require fire in the mornings and evenings. Our average of snow is 22 Inches covering the ground as many days in the winter. The necessaries of life are extremely cheap, but dry goods and groceries excessively dear, which renders it prudent to draw them directly from Philadelphia, New York or Boston, as they come to our doors by water. Our religions Presbyterian, Methodist, Baptist and a few Anglicans, a preacher of some of these sects officiating in Charlottesville every Sabbath. Our society is neither scientific nor splendid, but independent, hospitable, correct and neighborly. But the Professors of the University will of themselves compose a scientific society. They will be removable only by a vote of two thirds of the visitors, and when you are told that the visitors are Mr. Madison, President Monroe and myself all known to you by character, Senator Cabell, General Cocke, Mr. Watson, gentlemen of distinguished worth and information, you will be sensible that the tenure is in fact for life. Now, Sir, for the object of all this detail. I have stated that when men of the 1st order of science in their line can be found in our own country, we shall give them a willing preference. We are satisfied that we can get from no country a professor of higher qualifications than yourself for our Mathematical department, and we entertain the hope and great anxiety that you will accept of it. The house for that Professorship will be ready at mid summer next or soon after, when we should wish that school to be opened. I know the prejudices of every state against the climates of all those South of itself: but I know also that the candid traveler advancing Southwardly, to a certain degree at least, sees that they are mere prejudices, and that the real advantages of climate are in the middle and temperate states, and when especially above their tidewaters.

I must add that all this is written on the hypothesis that the legislature will confirm the report of the Commissioners. but that is not doubted: and therefore I make this early application to pray you to take this proposition into consideration; and as soon as you can settle your mind on it, to favor me with a line on the subject, shortening my anxiety for its reception only according to your convenience. In the meantime accept the assurance of my great esteem and respect.

Th. Jefferson

Nathaniel Bowditch Papers, Boston Public Library

Bowditch to Jefferson

Salem Nov. 4, 1818

Respected and Dear Sir

I have just received your much esteemed letter of the 26th ultimate containing the highly honorable proposal relative to the Professorship of Mathematics in the Central University, a situation which could be very pleasant to any one whose *engagements* would permit him to accept the proposal; but *several* important trusts (amounting to nearly half a million of dollars) undertaken for the children of an ancient merchant late of this Town, and which upon principles of honor cannot be resigned but with my life, bind me to the powerful hands of *interest* to the place of my nativity, endeared to me by the many civilities I received from my Townsmen while rising into life without any other advantages than such as the industry of a young person, thrown upon the world to provide for himself at the early age of 11 or 12 years, could procure. My income from these various trusts etc. exceeds $3000. Considerable time is left to me to devote to those studies which have been the delight of my leisure hours. My wife, whose health is very delicate is averse to any change of place which would remove her from those dear friends who have done so much for her in sickness. These and other circumstances of a similar nature must be my excuse for declining the professorship [offered] in your letter in such terms of kindness and civility.

Please to present to the Gentlemen Visitors, who have done me this honor my best wishes for them personally and for the success of the institution entrusted to their care, with the hope that the Professorship will be filled by persons eminently qualified to render the University highly respectable and beneficial to our country.

I remain Sir, with sentiments of high consideration and respect

Your most obedient, humble servant

N. Bowditch

Jefferson Papers, Library of Congress

A. M. Legendre to D. B. Warden[18]

Paris 2nd Feb. 1830

Sir,

There has been delivered to my house a card in your name with a very handsome copy of a 1st Vol. of the translation into English of the *Mécanique céleste* of Laplace, explained and commented upon by Mr. Nathaniel Bowditch of Boston. A very polite letter from the author accompanied this present, and I find therein a question relative to one of my works with which Mr. Bowditch is acquainted. In order to reply in a suitable manner to this letter, I intend to send to the author a copy of my work entitled *Traité des fonctions elliptiques,* 2 vols. in 4^0 and a 3rd containing some supplements of which two only have been published. Could you, Sir, obtain for me the occasion of passing along to the learned Mr. Bowditch the 3 vols. I speak of, with a letter of shipment. If your reply is favorable I shall have the honor of having delivered to you a package containing these objects, requesting you to confide them to a person who will be responsible for them.

I observe with great pleasure that the mathematical sciences, in their most elevated branches, are cultivated in the United States of America, with so marked a success, and I am particularly astonished at the magnificence of the typographical execution, a magnificence which exists to this degree only in the presses of England, for in this regard France is very far from being able to compete.

Accept, Sir, the assurance of the consideration with which I have the honor to be

Your very humble servant,
[Adrien Marie] Legendre

Legendre to Bowditch[19]

Paris 4 Feb. 1830 Quai Voltaire

Sir,

I have received the copy which you have kindly addressed to me of the 1st vol. of your English translation of the *Celestial Mechanics* of Laplace. Although I have reached the age where one is no longer capable of long and sustained application, it will give me truly great pleasure to read with all the attention of which I am capable the

[18] Translation of the original.
[19] Translation of the original.

notes and explanations which you have added to this work in order to facilitate its comprehension. What I have already seen does not leave any doubt that the great work that you have undertaken on this difficult subject gives you the right to the gratitude of scholars of every nation and especially to that of your compatriots who appear to have made great progress in the sciences, since you judge them deserving of reading and to getting to know works of so elevated a class.

You asked me Sir, if I have added anything to the 3 vols. which you know of my *Exercises in Integral Calculus?* I had the project several years ago of putting out a 2nd edition but instead I extracted the principal part and published it with many enlargements under the title of *Traité des fonctions elliptiques,* in two vols. in 4^0 with a 3rd vol. of supplements of which only two have yet been printed. These supplements are relative to very remarkable discoveries made in the same theory by two young geometrists of extraordinary talent, Mr. [Niels H.] Abel of Christiana, who has just been taken away from us by a premature death, and Mr. [C. G.] Jacobi[20] of Königsberg who promises to have a long and brilliant scientific career. I have addressed myself to Mr. Warden, your former Consul, to ask him to have sent to you by the 1st opportunity a copy of the work I have just spoken of. Please, Sir, accept this mark of my homage as well as my most distinguished consideration.

Legendre

R. D. Mussey to Bowditch

Paris Feb. 4th 1830

My dear Sir,

I should have written agreeably to your request from New York but I had not a moment to spare before embarking. I had a long passage of 36 days with some rough weather and head winds for the last 25 days. I arrived in this city on the 9th of January, and as soon as I was well settled, I distributed the numbers of the translation of La Place. Through the aid and kindness of D. B. Warden Esq. former Consul of the U. S. for this City. The Baron Zach[21] is in Paris. Mr. W. called at his lodgings with me but he was not in. We also

[20] Abel (1802–29) and Jacobi (1804–51) brilliant young mathematicians. Note that Bowditch was accepted as an equal and was learning about the latest work in the field fairly quickly.

[21] Baron Francis Xavier Zach (1754–1832) an astronomer, head of the Prague Observatory.

called upon the Marquis La Place, who is a Col. in the Army. He is a pleasant good natured looking man, and as Mr. B. informs me, is not at all distinguished as a mathematician. He appeared to be much pleased with the present, and said that he wants to ascertain and inform me as soon as possible at what expense 500 copies of his father's portrait might be had, at the same time remarking, he had met with a good deal of difficulty in procuring such a likeness for the work in French, as was satisfactory to him. Mr. Warden tells me that your work has excited a great deal of interest with those to whom it has been presented. He says that the gentlemen will soon make their acknowledgements to you with the exception of Mons. [Siméon Denis] Poisson[22] who, he says, is very lazy, and who promises to write on the receipt of the 2nd Volume. Mr. Warden took me to a session of the Royal Academy of which he is a member; and he invites me to go at any time when I please. After I become a little better acquainted with this outlandish dialect, I intend to go often, and perhaps I may, in a future letter, say something about the personal appearance of the mighty men of Science which compose that body.

Mr. Warden has a great deal of high American feeling, and he has had not a little gratified, in being able to present, to the Savans of France, such a work as yours. He told me that he should give some notice of it, shortly, in one of the journals. Mr. W. is a gentleman of extensive information, and is universally respected here. He requested me to ask you to do him the favour, if it would not give you too much trouble, to send him the result of the discoveries, made by the Nantucket and other American Whalemen in the Antarctic Ocean. He supposes that the required information is deposited in the Salem Museum; and is of Opinion that the principal important discoveries (commercial discoveries) were made by Americans, says he has had a controversy with an Englishman on this point. If you can furnish the necessary information, I sincerely wish you to do it. And as Mr. B. has a large library, and a very complete collection of American works, may I not suggest, entirely of my own accord, whether a present to him of a copy of your work, might not on the whole, promote the interests of American science. I assure you that you have a warm friend in Mr. W. that he is personally unknown to you. My sincere respects to Mrs. B. and family. Sincerely yours

R. D. Mussey

The address of Mr. W. is—Mons. D. B. Warden, Ancien Consul

[22] Poisson (1781–1843) was a leading mathematician of the day.

des Etats Unis, et Membre Cor. de L'Institut Royal—Paris Rue de pot de fer no. 12.

Accompanying this are two pamphlets from Mr. W. for Harvard College.

Marquis de Laplace to Bowditch[23]

Paris Feb. 10, 1830

Sir,

I have received with a profound feeling of gratitude the translation of the 1st vol. of the *Celestial Mechanics,* which Mr. Warden has been so kind as to send me on your behalf. It is pleasing to me to think that the work to which my father devoted so large a part of his researches has reached a country whose admirable development in that which truly honors the human mind he followed with so much interest.

The task which you have undertaken is long and not without difficulty, especially with the amplifications of which you are adding to the text in order to place it within the grasp of a larger number of people. I wish most sincerely and keenly that you may be able to conclude this vast work.

There is another of his works to which my father attached a very special importance which was for him a constant object of curious and profound meditation, even in the last years of his life; this is the *Theory of probabilities,* in which he used resources of analysis to solve a host of questions, hitherto the domain of philosophy and political economy. I was fortunate enough to follow him in these last works and to be associated in some measure with his thought. I know and can say how much he desired to see grow a class of knowledge which seemed to him useful in many connections, whether it be for exercising surely and for enlightening judgment, whether it be for dissipating prejudices, against which the best minds often cannot protect themselves; be it for evaluating the approximations of astronomical results and further as a new proof of that power of analysis which has served so marvelously as a lever to the genius of man in arriving at the most sublime discoveries.

If later on, Sir, your leisure should lead you to turn your attention to this new task and if you who feel the whole worth of the useful verities should judge it appropriate to produce on your own a work which will throw into relief some of the difficulties of the calculus,

[23] Translation of the original.

I stand ready to furnish to you with great pleasure, all the documents which are at my disposal. I do not know Sir, if you have the latest supplements that my father had added at the end of the *Celestial Mechanics;* there is one which we found in order among his papers and published about a year ago. If you should lack this, or any other, I shall hasten to send it to you.

M. Poisson asked me to present to you all his thanks for the copy which you have had addressed to him and of which he has as yet read only the notes.

Please, Sir, accept the expression of the sentiment of very particular esteem and high consideration with which I have the honor to be your very humble and obedient servant.

M[s] de Laplace

Bowditch Papers, Boston Public Library

THE WORLD OF THE NATURAL HISTORIANS

The United States was a natural historian's delight with its many species of plants and animals undescribed and unclassified. Rock formations provided interesting examples of geological phenomena and within the earth's strata were a new world of fossils. The accident of geography gave the natural historian a great advantage over the natural philosopher in America—he could learn from a bountiful nature near at hand.

With a scientifically unknown continent at their disposal, nineteenth-century Americans in great numbers collected flora, fauna, and rocks. The crowds at the popular lectures in chemistry were rivalled only by the size of the audiences for lecturers in natural history during most of the nineteenth century. But many more Americans became devotees and practitioners of the latter which remained a haven for amateurs and quasi-professionals for decades. Merely by keeping his eyes open a man could acquire a respectable reputation as a natural historian.

Yet even in the first half of the century the professional scientists in natural history, such as the botanist Asa Gray, were far removed in many respects from the amateur collector. The classification of species was not without its sophistications and abstruse points. Even more important, the best of the professional natural historians by the middle decades of the century were caught up in two trends which would alter their relationship to their amateur brethren.

Evolution drastically altered the climate of opinion and with it the nature of the taxonomic process. The timeless hierarchical relations of the earlier nonevolutionary taxonomic schemes became a net of genetic relations. At the same time the increasing use of laboratory methods, especially microscopy and physiological experimentation, would crowd the taxonomist from the center of the stage in the biological sciences. Taxonomy is not fashionable in these days of the triumphs of laboratory work. Yet the labors of the collector-taxonomist serves one preeminent purpose. By disclosing the diversities in nature they provide a challenge and a stimulus to the development of explanatory theories.

Natural history had two phases: collecting specimens in the field and the classification of the specimens in the study. The practitioners of these sciences, amateur and professional, constituted a community in which specimens were gathered, exchanged, transmitted, bought, and sold. To prepare his descriptions and especially to publish a significant classification a natural historian needed specimens and, by hook or by crook, they were obtained. Since there was often strong feelings that specimens might prove crucial coupled with the urge for first publication of a discovery, harsh rivalries and personal enmities were frequent.

The leading natural historians strived to form a network of relations with other professionals and with amateurs, often reaching across the Atlantic, which would yield a steady supply of specimens of plants, rocks, animals, as well as aboriginal artifacts. These were accumulated in "cabinets," often in conjunction with "philosophical apparatus" from the physical sciences. The development of the cabinets, the study of their contents, the use of the contents in classrooms and in popular lectures was the life of science for many Americans for well into the century. Practicing scientists were quite frequently annoyed, if not outraged, that the cabinets so often provided materials for display, not research.

Another characteristic of natural history differentiating it from both natural philosophy and the present day biological sciences was the often conscious linkage with art or esthetic purpose. We regard Audubon mainly as a great painter of birds; our predecessors proudly listed him as a great scientist. Feelings of wonder and awe at the marvels of nature were often tinctured with appreciation of their beauty. Since the accurate depiction of these marvels in the pre-photography era required artistic ability, the results were sometimes a work of art, or at least pleasing to the eye. Scientific accuracy and esthetic appreciation went hand in hand.

As the century progressed, scientific requirements were increasingly divorced from deliberate esthetic purpose in the biological sciences. As the scientists became more immersed in their laboratory recreations, amateur nature lovers increasingly served as the perpetuators of this aspect of the natural history tradition. For example, in the conservation movement of the early twentieth century scientists, as scientists, would largely support preservation of wild areas to have living examples of flora and fauna in a particular environment, not because of any belief in the intrinsic value of preservation of nature itself. The amateur naturalists frequently cherished wild beauty for its own sake. The biological scientist struggles with nature in the laboratory; the amateur naturalist passively contemplates the beautiful diversity and harmony in nature.

Audubon also exemplifies another attribute of the early natural historians. He published and largely acted as the salesman for his own works. Contemporaries of his were in somewhat analogous positions—they were scientist-entrepreneurs. In this respect they were lagging behind the physical sciences which increasingly relied upon organizations (learned societies, professional groups, government bureaus, and universities) to finance publications. Part of the reason for the lag was that the monograph, rather than the journal article, persisted so long as the vehicle for communication of research results in natural history. As the century progressed, natural historians of necessity abandoned their role as scientists-entrepreneurs and sought the aid of what would soon become the scientific establishment.

At the same time the business of getting specimens was becoming too complex for individual efforts, even when aided by a network of co-operating collectors. Audubon was a one-man ornitological operation; William Maclure, whose letters appear below, performed the virtuoso feat of a one-man geological survey of the United States (actually only the lands east of the Mississippi River). But knowledge grew too quickly, distances were great, and costs mounted. Although individual, heroic efforts did not cease, more often scientific work in natural history was conducted under the aegis of scientific societies, the national and state governments, and (laggardly) educational institutions. Succeeding the scientist-entrepreneur was the scientist-administrator, sometimes an active investigator but usually an intermediary between the researcher and those holding the power of the purse. Even in the early letters that follow signs of this trend are evident.

Prophets of a New Order

After accumulating a comfortable fortune, William Maclure (1763–1840), a London merchant, retired to devote his life to science and to the advancement of education, eventually settling in the United States and becoming an American citizen. Geology was Maclure's favorite field; his *Observations on the geology of the United States of America; with some remarks on the effect produced on the nature and fertility of soils by the decomposition of the different classes of rocks; and an application to the fertility of every state in the Union in reference to the accompanying geological map,* published in 1809, was the first national geologic map. His eminence as a scientist and his generous support of research earned him the Presidency of the Academy of Natural Sciences in Philadelphia, a position he held from 1817 until his death. Maclure had a unique role not likely to be repeated; he was a one-man scientist-foundation, conducting his own research and financing the work of others. When the support for science in the United States was meager and uncertain, the appearance of such a personage was a great event.

But Maclure is significant historically because he illustrates how the advancement of science influenced some individuals' aspirations for a betterment of the human condition. Looking around him, Maclure saw a world filled with ignorance, poverty, and injustice. Most of mankind were condemned to drudgery and to pain for the benefit of a few. The world of men was certainly not running very well. But in contrast, the world of nature disclosed by science was a model one where no disharmonious inequities existed. Could not the world of men attain the same perfection? After Newton's successful exposition of the mechanics of the universe, many men answered that question in the positive. And science was to point the way.

Maclure's approach to the problem was by no means unique. He saw in education the means for achieving a new order among men. As in the case of his scientific work, Maclure was unusual in his role as a source of financial support for the carrying out of his educational projects as well as those of others. No chasm divided his scientific activities from his educational endeavors. They were all of one piece; training in science (which was useful) replaced the useless teaching of the past. The content of conventional education was deliberately designed to prevent mankind from dealing effectively with their environments. Only knowledge of the pure and applied sciences could enable man to replace the old horrors.

Adults were beyond teaching in his view; only the young were not yet corrupted by the world they lived in. Early in his reforming attempts Maclure discovered that even the young resisted his earnest efforts because of their parents' influence. A lesser man might have despaired after this discovery, but not Maclure, who finally decided to educate orphans according to his system. Besides orphans Maclure needed a group of research scientists to carry out his educational experiment. The students were to learn science by working with men actually engaged in research.

His most ambitious educational and scientific project involved Robert Owen's New Harmony community in Indiana. Owen (1771–1855) was a Scotsman who had made money in manufacturing and like Maclure was promoting unorthodox beliefs. Unlike Maclure, Owen believed in the possibility of founding viable utopian communities where property was held in common. For the sake of the opportunity of educating the New Harmony children (at first those with parents) and of establishing a scientific press and research group, Maclure joined Owen.

In 1825 the keelboat *Philanthropist* left Pittsburgh to go down the Ohio and up the Wabash. It was a "boatload of knowledge" containing among others Maclure; the entomologist Thomas Say (1787–1834); Charles Alexandre Lesueur (1778–1846), the French zoologist; and Gerard Troost (1776–1850), the Dutch geologist. For nearly a month the *Philanthropist* was caught in ice in the Ohio, but the party eventually reached its destination. Joining them later was Cornelius Tiebout (1773–1830), an engraver whose skills were necessary for the scientific publications. Maclure and Owen soon fell out, and Maclure went to Mexico in 1828 to continue his educational ventures. But New Harmony, a small town in a frontier state, continued to produce scientific work and to educate the citizenry. Maclure remained a generous patron of American science in New Harmony and elsewhere until his death.

Maclure was an old man when he wrote the letters that follow. No longer active in science, he still ardently pursued his schemes for a scientifically educated citizenry. The crotchets of an old man are evident, perhaps even signs of senility. His scientific work was being superseded by the work of others who extended their geological investigations beyond the surface level that Maclure had studied. To the modern eye the views he espoused are a strange mixture of the acceptable, the irrelevant, and the radical (such as his views on property). Maclure still inspires respect as a man of ability attempting great and good things. He was the father of American geology, the most extensively developed scientific field in nineteenth-century America. At a time when science was honored but very modestly supported, he attempted a program in research and in education far in advance of his era. Even though his indignation at the wrongs in the world did not produce overnight changes, Maclure's honesty and sincerity merit respect.

Faced with men like Maclure—radical reformers—both conservative ire and popular opinion often jumped to the conclusion that these individuals were typical of scientists. Maclure's views were certainly not shared by the majority of his scientific contemporaries. The American scientific community of the past century was a heterogeneous group, and it is not possible to give a detailed, definitive exposition of their nonscientific beliefs. But for every Maclure it is not difficult to point out at least one scientist of unimpeachable orthodoxy. There was, at the most, a greater tendency towards liberal views among the scientists than in the public at large, but even that generalization is a shaky one.

As specialized scientific disciplines developed, a notable trend appears. There is a diminution of concern with the atypical beliefs and public activities that attracted some of the early natural historians and natural philosophers in the days when the line between professional and amateur was blurred. Most of the professional scientists of the late nineteenth century and the early decades of this century confined their innovating within the boundaries of their discipline. Many were quite conservative or had rather naïve flirtations with heterodox views. The true heirs of Maclure and his kind were the reform-minded social scientists not always held in high esteem by their colleagues in the natural and physical sciences.

Thomas Say to Benjamin Tappan[1]

New Harmony August 30th 1827

Dr. Sir.

Your letter of the 29th ultimate was duly received. Dr. Troost has left here some time since. I believe him to be a mineralogist and chrystalographer of the first class, but did not expect that he would publish largely, or give to the science any extensive work and I do not suppose that he had any expectation of it himself. As to Mr. Maclure's plan of publishing books, he has always said, and yet says, that even should his other intentions be successfully counteracted,

[1] Tappan (1773–1857) was a member of a notable family of philanthropists and abolitionists. As Senator from Ohio, 1839–45, Tappan was influential in passing various legislative enactments involving science, including the bill on the Smithsonian bequest.

the business of Science shall go on. With respect to the Entomology, we have offered to Mr. [Samuel Latham] Mitchell,[2] proprietor and publisher of the work, to proceed immediately with it; that the plates should be engraved printed and coloured here, as faithfully as those done in Philadelphia and that we should pay the expense of transportation. This proposition is so advantageous to him, that we cannot suppose he will reject it; but we have not yet heard from him on the subject. Two Nos. of Mr. Lesueur's work on Fish with coloured plates have already been published, and the third No. is nearly ready to be printed off. The first plate is now engraving for my work on Shells. So that you see we are at least making something of a spirited commencement as publishers.

Mr. Maclure, in a letter written some time since, accounted to you for the money, which you left with me, for the education of your children, the Balance, $77 and some cents, was paid to Mrs. Neef. You will please destroy the receipt I gave you.

If Mr. Lesueur should incline to leave Harmony, I shall certainly lay before him your very liberal and handsome offer, but he has often said to me, that if he should leave this place, he would go to France.

Mr. Maclure's reasons for his change of plan, are the following: he came here from his confidence in the judgement and integrity of Mr. Owen and hoping to find more toleration for introducing his plan of education, under the community system, in all of which he has been grossly deceived, finding that the aver[s]ion to the system of education was 100 fold augmented by the unpopularity of Owen's system; and finding himself too old to contend any longer against the ignorance and deep rooted prejudice of parents he gives up all attempts at giving a rational education to those who can afford to pay for the education of their children, and returns to his original plan of giving knowledge cheap to the poor and oppressed. It has always been his intention to diffuse knowledge amongst the great class of productive labourers; this was the only object that induced him to meddle with education, being thoroughly convinced, by all his experience that the inequality of knowledge is the source of all the evils that torment humanity, and giving knowledge to the rich is the certain way of increasing that inequality: which, in other words is increasing the miseries of our species. The cost of ten month's

[2] Mitchell's journal, the *Medical Repository* is sometimes described as America's first scientific periodical.

experiment on two hundred unproductive people, the school community, was also a discouraging reason.

Mr. Maclure presents you his compliments and my wife wishes to be remembered to Mrs. Tappan and believe me to be as usual, with great esteem,

Your Obedient Servant
Thomas Say

P.S. I copy, below, Mr. Lesueur's prospectus; if you can obtain any subscribers for him he will be much benefitted thereby.

Proposals for publishing by subscription, a work on the Fish of North America, with plates, drawn and coloured from nature. by C. A. Lesueur.

This work will be published at New Harmony, Ind., in numbers, with four coloured plates in each, and the necessary letter press containing the descriptions of the species represented. Twelve numbers will constitute a volume. Messrs. Tiebout and other artists from Philadelphia who were there engaged on the "American Entomology" are occupied on this work. Books with coloured plates, are generally beyond the reach of persons of limited means; but it is intended, that the present work shall be adapted to the circumstances of all. The price to Subscribers will therefore be forty cents each number.

Benjamin Tappan Papers, Library of Congress

William Maclure to Samuel George Morton[3]

Mexico April 3, 1830

Dear Sir

Your letters of the 10 August and 12 September 1829 only just received owing to the box in which they were being detained at Vera Cruz, agreeable to the negligent mode of doing small matters in this country, merchants being but bad scientific agents, I received at same time two boxes of books and papers that had been detained 9 months at New Orleans. We find it very difficult to find any agent at any of our large towns who will pay the smallest attention to our small affairs for our school; whether from mere negligence or an invertebrate antipathy at every thing that can make knowledge cheap

[3] Morton (1799–1851) was an anatomist. He is best known for his study of the human skull. As secretary of the Academy of Natural Sciences, he corresponded much with Maclure, the Academy's wandering president.

to the millions. As you may have seen by our school sheet, the disseminator (which I ordered to be sent to the academy) the sole object of our establishment at New Harmony is the equalization of property, knowledge, and power—the *sine qua non* of human happiness. The first and last are out of the reach of individual exertion. Knowledge is the only one that individual efforts can have any effect on.

It is with that view that the remainder of my short existence and property is devoted. I thank you for your account of the prosperity of the academy it affords me much pleasure. Having only one correspondent at Philadelphia that could give me any account; and he is so occupied with things he thinks of greater consequence that I seldom hear from him. Your exertions have been exceedingly useful and your plans of further utility will be still more. I am only sorry that my friends will not permit me to aid you, as I have not half the revenue I had when I was enabled to do something for the academy. Having embarked $60,000 in endeavoring to form an establishment of industry in Spain and $90,000 sunk in an unproductive property at New Harmony in consequence of attempting to aid Mr. Owen in his community system, which last may perhaps become in time the support of the school of industry, tho at present it gives little or nothing.

I am grown old and all my joints a little stiffer than they were. Most physical exercises ceases to be a pleasure and therefore the practical pursuit of natural history must be left to younger limbs. The progress that we have made in the knowledge of all the exact sciences is really astonishing when I compare the present with the past. When a small cabinet of mineralogy which Dr. [Adam] Seybert[4] brought from Göttingen which cost him 15 dollars was the only one we had to run to in Philadelphia no more than 30 years ago. The progress will be geometric 30 years hence the most enthusiastic conjecturer can scarce divine the state of the arts and sciences, and every other improvement and inclination in favor of humanity. On this side of the Atlantic is the slavish imitation in all moral and speculative matters such as religion and politics, of old Europe, for some time past in her dottage—and only lately begun to revive, for even the aristocracy of the old world, after 1000 years of uncontrolled dominion, are forced by the consequences of ours and the French revolution to permit the millions to enjoy a little more of

[4] Seybert (1773–1825) was a physician who did much research in chemistry. Seybert also served in Congress.

their own property than has yet been allowed them. And should the stand our operatives seem willing to make succeed, it will no doubt, like the principles of the theory of our and the French revolution, spread rapidly. I say theory for it's only the theories of freedom that has yet been enjoyed by the mass of mankind in any country; the practice has been so prevented by the ignorance of the many and cunning of the few, that the practice of freedom and equality is even with us I fear some ages behind the theory. It's therefore not to be wondered at in this half barbarous country emerged only yesterday from a state of foreign slavery, that the elements of slavery and tyranny should yet predominate and deceive the young short sighted politicians who never experienced our state of poverty, and total privation of every comfort and convenience during and for some time after our revolution, and boldly predict that this people can never be free. That they are so ignorant at least to all self consideration etc., that they are only made to be slaves. In all of which I differ from those superficial observers, as I have never been in any country where the practice of equality amongst all ranks, was so evident, the native Indians the 4/5 of the population are deficient in none of the ingredients of civilisation, have quick capabilities of imitation, the habits of industry and tho at present ignorant yet they know it and are convinced that their ignorance is the source of all their troubles and miseries. A knowledge that the pride and presumption of those who consider themselves vastly their superiors has prevented them from acquiring. All who are convinced of the incalculable evils, troubles, and miseries of ignorance are full half way on the road to useful knowledge, and all they acquired are useful because put into immediate practice and not like [90%] of a classical education consisting of sounds applicable to none of the purposes of life, and forgot the moment we leave school, but from the tedious purgatory we passed thro', frightens us from study during our lives, preventing the daily practical improvement, the consequence of being taught pleasantly only the useful.

All my amusements are chiefly moral and the gratification of observing a rapid course towards improvement one of my principal pleasures, which is fully satisfied in impartially watching the great progress made daily by the millions in this country, for the ruling few are perhaps at a stand if not retrograde . . . The sudden transfer from oppression to freedom has so intoxicated them that the old proverb "set a beggar on horseback" is revived; but the industri[als] we produce are gaining knowledge every day, and are only waiting a convenient time to claim their rights of universal suffrage when they will be almost as far advanced as we are. This may appear a new doctrine and like everything new against all that are well off in the old order of things. A Mr. Grimes from London has been occupied for the last fort night in teaching 8 or 10 soldiers to read Spanish by a new and simple method he gives them a lesson every working day of 1½ hours and already the best capacities begin to read and tho some of them are dull and others for want of confidence not attentive yet there is no doubt they will all read in a few weeks more which fully proves the practicability of the system, which is something like what Lafevre invented in France to teach in from 9 to 40 hours to read any language according to the capacity of the learner by this method all the Indians both male and female may be taught to read and also to write in a few months, as after a certain time more are to have the right of suffrage but those that can read a stimulant that will enduce them all to learn. I remain yours sincerely

Wm. Maclure

Mexico 3 April 1830

Dr. Morton Philadelphia

P.S. Mr. Say and Lesueur are endeavoring to propagate a taste for natural history in the west through the medium of our free and economical press wrought by the pupils of our school of industry it goes heavy like all new things having the opposition of all who benefit by the old. The moral revolution that is only just begun by throwing more power into the hands of the producing classes will equally augment their property and knowledge which will be in time to make the cheap editions of useful books our situation at New Harmony will enable us to publish. For soon the old books on Law, theology, metaphysics, of history, and literature will be obsolete and must be replaced by something more assoc[iated] with the knowledge of the day. A vast advantage these young candidates for civilisation have their libraries tho small have nothing but the useful which the speculators in Europe have furnished them all translated into Spanish offering correctives for their moral and physical errors. All the novelty in politics and religion that has ever been printed in any country or language an assortment many of which none of our booksellers would like to put on their counters are to be found in Spanish in all the shops. Their civilisation will be grafted on the most useful to be found without the incumbrance either moral or physical of the vast rubbish accumulated in the various unsuccessful attempts at improvement an advantage we have benefited by, but they will profit still more in proportion to the higher state of civilisation at which they commence their career there are all novelties not to be found in ancient or modern history like the properties of federation not known to those who have been enjoying its advantages for the last 53 years in but an imperfect state like all new experiments which I hope we will have the good sense to reform and bring it nearer to the knowledge of the day.

Say to Morton

New Harmony March 5th, 1833

Dr. Sir:

I have great cause to apologize to you for this long delay in acknowledging the reception of your polite and acceptable letter; but the hope of being able to express to you my thanks on the reception of those desired fossil shells, which you were so good as to destine for me, contributed chiefly to this apparent neglect of courtesy. For the fate of that collection of shells I should entertain the most lively apprehensions, did I not know, from my own experience, that packages can be obtained in a much shorter time from Europe than from Philadelphia. I therefore yet have hope that the collection may arrive safely, as I have hardly a specimen from Maryland or New Jersey of a fossil shell; and if so, I should be gratified to increase the interest of my work by inserting any geological observations relative to those locales, which you may find convenient or agreeable to furnish me with, in your own name. The publication of my work has been retarded by the protracted illness of our engraver; but on his recovery I hope to prosecute it with some energy and I am very desirous to make it as complete as possible; with this view I solicit the aid of all my friends, for specimens and observations, which I shall in every case, be particular to acknowledge.

I regret extremely to have given the society so much trouble about my Mss. It was not my intention to excite their attention to the subject at all, and I beg to return to them my thanks for their exertions to ascertain the fate of these unfortunate descriptions. In a note to your esteemed letter you gave me the following notice "If I understood Dr. Pickering[5] rightly, descriptions of 3 or 4 insects, which were insufficient to complete a *form,* were omitted by the

[5] Charles Pickering (1805–78) a naturalist, later with the Wilkes Expedition.

printer, respecting these I will make further enquiry." Now these are the very descriptions, I make no doubt, which I have been so clamorous about for so long a time, and, I fear, worrying out the patience of my friends; excepting, that instead of 3 or 4, I think there ought to be from 8 to 12 of them, I feel the anxiety which I could not avoid expressing these few past years. But a truce with so disagreeable a subject of "Still carping upon my daughter."

As soon as I shall be able to collect sufficiently good specimens of our Wabash and land shells, which will be, no doubt, the next low water of our rivers, I will do my best to repay your liberal present of fossil and other shells, though I have hardly a hope to have in my power a complete remuneration.

I have been for a long time preparing a "Species" of our Mollusca, as well as another of our Insects, both of which I hope to publish, but I fear I shall be under the necessity of examining our coasts myself, to ascertain the geographical distribution of the former as well as their manners and habits etc. And also the various localities of fossil shells, so numerous in our country.

Very Respectfully yours

Thomas Say

Lest the omission may lead to error I will state that I have other missing Mss. in Philadelphia which I am more anxious to obtain, than the above and which I have been equally vociferous to obtain, as the want of them, or of knowing what they are, has for three years prevented me from publishing an essay on the same order, containing about 150 new species; they are the continuation of "Descriptions of the new species of Hymenoptera" in the "Contributions of the Maclurian Lyceum" No. 3. where on page 83 the essay is said "(To be continued)" but it has never been, and the Mss. has not yet been given up to me, and I am at a loss what further steps to take in the truly disagreeable business. Did the Academy of Natural Sciences and Philosophical Society ever receive my "Descriptions of new species of Curculionites of North America" and "Descriptions of new species of Heteropterous Hemiptera of North America." etc.

T. S.

Morton Papers, American Philosophical Society

The Botanists' World

One attribute of science is its internationalism. Research enterprises of any significance are not confinable within national boundaries. Botany provides a particularly good illustration of this tendency. Since the time of the great Swedish botanist Linnaeus (1707–78) a lively international network had linked scientists in this field. Plants were exchanged and sold on a world-wide basis. Both amateurs and professionals were involved; as the field developed, the former would become less significant.

Identification and classification of the plants was the principal object of the international network. To properly accomplish this, a botanist needed not only reference works giving past descriptions of plants but also a herbaria containing preserved specimens. At early stages in the development of botany the amateurs and professionals might have a rough parity in these necessary working arrangements. Increasingly, elaborate institutions were required, and the solitary amateur was at a disadvantage in collecting and in classifying. Detailed knowledge of the past investigations was essential both for gathering what was known and for properly placing a newly found species in the taxonomic system.

Consciousness of the past eventually created a kind of scholastic tradition in botany and, to a lesser extent, in other taxonomic efforts. Good science and good manners both required that predecessors get full credit for discoveries. Naming of the species involved assignment of priority in discovery, often using the name of the scientist involved. A careful search of publications, an examination of old records, and a comparison of specimens were often involved. Many botanists went to inordinate lengths in checking priorities, meticulously crediting others. When one of their colleagues neglected to cite his predecessors, such a lapse in botanical manners was quickly and often acidly pointed out.

The letters of two dissimilar individuals, perhaps representatives of different stages in the development of a formal scientific discipline, appear in this section. Constantine Samuel Rafinesque (1783–1840) came to America first in 1802 and finally settled here in 1815. From 1818 to 1820 he taught at the then prosperous Transylvania University at Lexington, Kentucky. Afterwards he engaged in botanical excursions and eventually settled in Philadelphia, his latter years being spent in poverty. Later critics could not agree on the merits of his contributions and whether he was brilliant or merely eccentric. Part of the uncertainty stemmed from the former state of scientific publication in the United States. Rafinesque's writings are formidably scattered, not neatly clustered in conventional monographs and journals. Partly through his own peculiarities and through accidents beyond his control, his collections, notes, and other records were dispersed, some lost forever. Rafinesque was hard to judge because his work did not fully enter the main stream of botany.

John Torrey (1796–1873) was a physican who was a teacher of chemistry. At various times he occupied professorial chairs at West Point, the College of Physicians and Surgeons in New York, and the College of New Jersey (now Princeton University). After 1853 he was the Chief Assayer of the U. S. Assay Office in New York. Careful, neat, and diligent in his work with specimens, Torrey collected and named thousands of new species. The gathering and exchanging of specimens by professional botanists and amateur nature lovers was carefully nurtured by him. Torrey became, consequently, an important figure in the international network in botany. He encouraged, befriended, and stimulated young scientists. Some would follow him into botany. Of these the most notable was Asa Gray (1810–88), who appears in the letters that follow as a young physician seeking a place in science.

John Torrey to Constantine S. Rafinesque

New York December 9th 1833

My good friend.

I have been (as you may well suppose) fully occupied in various ways since my return from Europe. During an absence of six months much business accumulated which I had hardly disposed of when it was time to make preparation for my Chemical labours in the Medical College. I have snatched a few moments this evening to reply to your letter of the 12th ultimate and I must write briefly. As I propose visiting Philadelphia some time between the 25th and the 30th of this month (when our lectures are [intermitted]) I will not tell you much botanical news, for we can talk over these matters at our leisure. I will merely state that I spent more than a month in [Sir William Jackson] Hooker's family and, of course, saw much of that distinguished botanist and excellent man. He is very laborious and uncommonly accurate. His Flora bor. Amer. has reached its 6th No. and the 7th is proceeding. His botanical miscellany has now advanced to the 4th vol. and a most interesting work it is. He and Arnott have completed the 3rd No. of the Botany of Beechy's Voyage.[6] I was a week with Arnott on the shore of Lochlenen and another week at Edinburgh where I saw much of Dr. Graham, Regius

[6] Hooker (1785–1865) was the head of botanical gardens at Kew. He and G. A. W. Arnott (1799–1868) were preparing a description of the plants brought back from the Arctic by the British naval officer, Frederick William Beechey (1796–1856). Robert Graham (1786–1845) like so many others in his period was a physician originally. Robert Brown (1773–1858) was the heir to Banks' scientific collections which went to the British Museum with himself as curator. Brown was renowned for his studies of plant reproduction.

Prof. of Botany there. In London I was much with R[obert] Brown who is justly considered the greatest botanist of the age. He looked over many of my rare plants. I also became well acquainted with Lindley, Bentham, Cunningham, Booth, and other London Botanists. I looked over nearly all Persh's [i.e. Pursh's] plants in Lamberts Herbarium; they are in miserable condition. I had some pleasure in studying the Herbarium of Gronovius and by means of it I was enabled to settle some difficult points in our Botany. I remained in Paris only a month, but I saw much of the botanists there, particularly the younger Jussieu, A. Brogniart, Decaisne, Guillemin etc. I examined with great care the Herbarium Mich[au]x and took notes of every thing interesting that I observed. When you come to New York I will show you the books that I brought home and some of the rare North American plants from the North West and other parts of N. America presented to me by Dr. Hooker, Lindley and others.[7]

I received no letters from you while I was abroad. When your plants are arranged I should be glad to look over them. Perhaps you will have part of them ready by the time I arrive in Philadelphia. The later numbers of your Atlantic journal I have not seen. Indeed I think the 4th was the last that you sent me.

Mr. Schweinitz[8] found the plants of Baldwin's collection he purchased to be very valuable. He certainly got them a great bargain. I am to have some of the duplicates.

Yours truly
John Torrey

Torrey to Rafinesque

New York February 5th 1834

My good friend

I am almost ashamed to make you another promise, after having disappointed you twice. It was my full purpose to visit Philadelphia during the Holidays and I anticipated much pleasure in seeing the many friends that I have there. It was very important however for me to be in New York at the time that I had set apart for my visit so that I could only make a hurried excursion to Princeton where I

[7] Torrey is dropping the names of his scientific contemporaries except for the references to Pursh, Gronovius, and Michaux, whose specimens from North America he undoubtedly examined with care.

[8] Rev. Lewis David von Schweinitz (1780–1834) was an American Moravian clergyman who specialized in the study of fungi.

had some business to attend. By leave of Providence, I hope still to be with you before another month passes. I must see Mr. [Thomas] Nuttall[9] before he leaves Philadelphia. You know that he intends making the arduous journey across the Rocky mountains to the Pacific ocean. He will be gone two years, and I have but little doubt that he will collect a multitude of the most interesting plants. Have you seen the collection which he is now examining made by a friend of his during his journey this way from the Pacific? He proposes publishing an account of them in the journal of the Academy if he can finish his examination in time. I showed him the choice things that Dr. Hooker gave me from the North West and he found them of the greatest use in enabling him to recognize the described species. Mr. N. means to present all his exotic plants to the Academy, but I am to have the duplicates.

The little time that I could spare from my chemical duties has chiefly been devoted to the study of grasses and cyperaceae. I am very anxious to see your collection of these tribes and hope to inspect them next week. Be sure to have them all by themselves. My herbarium is very rich in these orders and I understand them now tolerably well. Did I tell you that my friend and assistant, Dr. [Asa] Gray was about publishing the 1st No. of his North American Gramineae and Cyperaceae, or dried specimens of these plants. Each number is to contain 100 good specimens fastened on handsome white paper bound in neat covers, with printed labels, title page, index etc. The price is 5 dollars per number. For the same specimens not fastened on paper, nor bound in a volume, the price will be 4 dollars the hundred. I think his prices are low. Dr. Gray will probably visit the state of Georgia next spring for the purpose of collecting plants, unless we can have him sent as a naturalist in the expedition that is to explore the country under orders of Government next Summer. I shall soon begin to work in earnest at my 2nd vol. and at the same time digest my copious materials for a synopsis of all North American plants north of Mexico. Hoping to see you soon when we will have a long talk about botanical matters. I remain

Yours truly

John Torrey

Rafinesque Collection, American Philosophical Society

[9] Thomas Nuttall (1786–1858), an Englishman, was an ornithologist as well as a botanist, and did collect much as predicted by Torrey on this trip.

Torrey to Asa Gray

Medical College, New York Saturday evening
March 1, 1834

Dear Doctor

I closed the Chemical Course last evening, and as you may suppose, I feel not a little relieved. Still I have an unpleasant task before me, the examination of the Candidates for the Degree M.D. By way of relieving the tedium I mean to scribble a letter to you, though I proposed writing to you as soon as the College business was concluded. With Mr. Clark's assistance, and a little extra exertion on my own part, I was able to make all necessary preparations for the lectures. The young men behaved very well to the last hour.

I hope you had a pleasant journey home, and that you found all your friends in good health. Here, we continue very much as you left us.

Within a week I have received some packages of plants

1. A choice collection from Prof. Fischer of St. Petersburgh, chiefly collected in Siberia, all neatly labelled, and Fischer's own name written on each paper! There are about four hundred plants in this lot, many of them species new to me.

2. A collection of very nice things from North Carolina, sent by the Modest Mr. Moses A. Curtis, whose letters have amused me not a little.

3. A good collection from Florida, the parcel alluded to by Lt. Alden, as sent in his box of books. I only received it yesterday!

I received a few days ago a letter from Dr. Loomis one of the compilers of the Newbern Catalogue, informing me that he had transmitted a small collection of plants to me principally species that I had marked in his list. The parcel is in town but it has not been called for yet.

Mr. H. B. Croom, the other author of the Catalogue has written to me from Tuscaloosa informing me that he had taken up his residence in Florida and would collect plants for me as far as his feeble health and professional duties (as a lawyer?) would permit.

One of our students, who returns soon to New Orleans, and has herborized with Geo. Clinton, volunteered to collect largely for me in Louisiana. I think he will not disappoint me.

I have received *duplicate* bills of lading of my plants etc. shipped for me at Havre so long since, but the vessel which has the box on board, (though long since due) has not yet arrived. I do not yet dispare, but "hope deferred etc."

Mr. [William] Oakes has written me a long letter. He is still working very leisurely at his New England Flora, but he will not publish in less than three years, not until I have come out with the synopsis. He complains, with good reason, of the great number of nominal species with which our Flora are crowded, and says we must not spare them. I do intend to cut them down without mercy and have already decimated two or three regiments.

LeConte was in the laboratory last week, he has received no answer to his enquiries respecting the Cavalry expedition. Neither has Maj. Delafield heard from Washington on the Subject.[10]

I have almost made up my mind to deliver a Course of popular botanical lectures in this city, previous to my going to Princeton. My friends say that I shall succeed, and that a choice class can be raised with very little exertion. I shall probably lecture at Niblo's Saloon, a very respectable place, and conveniently situated, for the rich people of the "West End." It *is said* that I can have the hall of Columbia College to lecture in. This would save me at least one hundred dollars, but I could hardly expect so large a class there, as at Niblo's. When I have completed my arrangements I will write to you again. If I do lecture in New York, I shall not leave the city for Princeton till the middle of May next.

Yours truly
John Torrey

P.S. Monday Morning. We are all well this morning, and are chatting as usual a few minutes after breakfast. The children often talk about you. They say that they do not have such pleasant rides on your back as they used to have. Mrs. T. desires me to send you her kind regards, and Edward wishes to be remembered to you, and says that he thinks you might have written to him. . . .

J.T.

Gray Papers, Harvard University

[10] John Eatton Leconte (1784–1860), an entomologist whose son John Lawrence (also an entomologist) and two nephews John and Joseph would be important members of the next generation of American scientists. John Delafield (1790–1875) was the president of the New York Lyceum of Natural History and was apparently trying to get Leconte a position with an Army exploring party in the West.

C. W. Short[11] to Rafinesque

Lexington, Ky. Sept. 7th 1834

Dear Sir.

Your communication of the 5th August came to hand some weeks ago; but as you mention your intention of being absent from Philadelphia until the 1st of Sept. I have delayed writing until now.

For the last three or four years I have been very industriously engaged in collecting and preserving the plants of Kentucky, as my leisure and opportunities afforded, during the Spring, Summer and Fall, the winter months, you know, I am confined to Lexington. During the last season I collected very largely, and put up and sent away to various correspondents in Europe and America about 5000 specimens of about 600 species, reserving still for my own Herbarium an abundant supply. This season I have not been less industrious, but have already collected a very considerable parcel, and hope still to add to it a large number of our autumnal plants. The Season has proved eminently auspicious to the preservation of plants, it having been universally dry, so that this year's collection is in fine order; and I have avoided the [two words unclear] hitherto used too indiscriminately in the preparation of my specimens. I say this much with the view of informing you what you may expect from me in the way of exchange. Should we enter upon one; but I have further to remark that it does not suit my convenience to distribute my collection into parcels before the winter; during the spring, summer and fall my whole leisure time is employed in traveling, collecting, examining, and preparing my plants. In the winter I devote my hours of leisure to the distribution of those which I have to spare, into parcels for my friends, and the arrangement of my own Herbarium. I have already engaged to make out during the ensuing winter parcels for Dr. Hooker; M. Michel; Dr. Booth; (of Europe) Dr. Darlington; Dr. Torrey; Mr. Curtis of N.C.; Mr. Croom, of Florida; Mr. Durand; Dr. Griffith; and the Am. Philos. Soc. of Philadelphia, Dr. Aiken of Maryland; Dr. Leavenworth, U.S. Army, at Fort Towson, Arkansas, and Mr. Riddell of Ohio. From several of these I have already received parcels in anticipation, and am, therefore, in honor bound to supply them;

[11] Short (1794–1863) was another botanizing physician. This letter, in no way different from many others of the period, is a typical example of the specimen-gathering and distributing process.

nevertheless I shall have more than sufficient for all these demands, from which I may be able to make out an interesting parcel for yourself. In the mean while, however, as you have the plants on hand, and leisure I hope, to put them up, I must desire of you an advance of one parcel, with the insurance that, during the winter, you shall be repaid with interest. In the selection of a suite from you I would especially wish to have specimens of all your North American discoveries, the new genera of species which you have discovered or framed on former discoveries; and especially of Western and Southern plants. In addition to these I should like much to have specimens foreign and oriental medicinal plants. A parcel of this kind made out as soon as your leisure permits, and placed in the hand of Mr. J. Dobson of your city will reach me safely.

I was certainly not apprized of your having published a catalogue of Kentucky plants, much less *two* of them; in that which we made out, and to which you refer, we did not pretend to give a perfect list, but one of those plants only which we had actually met with here, intended for the convenience of our correspondence in making out a list of their desiderata. Where are your Catalogues to be had, I should like much to possess these? The *Iresine cedorioides* grows abundantly on the margins of the Kentucky river at almost every point where I have visited it, preferring the moist sandy alluvions and shaded situations, flowering from July to October. The *Enslenia* is still more abundant and in similar situations I have never known it to fail flowering freely in July and August. . . .

Rafinesque Collection, American Philosophical Society

The Unity of Science

For most of his life Torrey supported himself and his family by teaching chemistry or the practice of chemistry in the Assay Office, not as a botanist. Many other scientists of that time still had active professional interests in two or more scientific fields, if not tendencies towards encyclopedic coverage of all sciences. In this section and in a later section on the physical sciences Torrey appears not merely as a botanist, but as a chemist-botanist. William Maclure, an old man awaiting death in Mexico at the time of the letters below, continued to write Morton, the patient corresponding secretary of the Academy of Natural Sciences keeping in touch with his distant president. Dividing his collection of specimens and his books then occupied Maclure's thoughts, as well as other provisions for the various ventures he founded or supported.

In these letters of Maclure another theme appears—the building of a proper astronomical observatory in America, for many years the fond dream of American supporters of science. This wide-ranging interest in many branches of science was typical of the day. When we now speak of the unity of the sciences, what we are referring to is largely a philosophical concept, not a description of the actual life of science. Even when two or more scientific disciplines converge on a common problem today, the normal tendency is not for a fusion of the original two but for the development of a new discipline. Things were simpler in the 1830's, and in the small scientific community of America scientists could soon know or have contact with many like-minded persons in the nation. And they were often interested professionally in the work in seemingly diverse fields.

Two other trends are visible in the letters: the appearance of men who would later greatly influence science in the United States and the rising tempo of organized scientific activities, especially the expeditions to the West and overseas. Asa Gray, who had found his life's work in botany, appears again in Torrey's letters as his older friend and colleague is concerned with finding Gray a suitable position. Joseph Henry, with whom Torrey corresponded about obtaining a post for Gray at Princeton, had already notable achievements in the study of electricity and magnetism to his credit. Still before him was his role as Secretary of the yet unfounded Smithsonian Institution and as the great patriarch of American science. Alexander Dallas Bache, mentioned by Torrey in his February 2, 1835, letter to Henry, was Professor of Natural Philosophy at the University of Pennsylvania. Before too long he would become the great scientific promoter, organizer, and administrator of his generation—the great tycoon of American science.

Torrey to Joseph Henry

College of Phys. and Surgs, New York
February 2nd 1835

My dear friend

Seated at the table in our Examination Hall, I feel that I cannot while away the time more agreeably than in writing you a few lines. Your letter of the 23rd ultimate came to hand on Saturday and this morning Mr. Hart (as I suppose) came into my laboratory to tell me that he had a parcel for you, enquiring whether I expected to have an opportunity of sending it to you at Princeton shortly. I sent for it at once, and it is now lying on the table at my side. I know not where to find your Mr. D. Hart, or I would send it by him. He dropped your letter to me into the postoffice. Dr. Rhinelander says that he will send to Princeton on Wednesday and perhaps I will entrust the parcel to him.

Do you mean that I shall not be idle next summer! You must not expect me to lecture much oftener than I did last summer. Not that I wish to be idle, but the young men must have time to study the subject, and to attend recitations. Pray let us have the text books in readiness this year. I will ascertain in time whether there will be a new edition of Linneaus out before the Course commences.

I wrote to [Alexander Dallas] Bache last week, and returned thanks for my election. It is not probable that I shall send them a paper very soon for I have work on my hands that is hardly the thing they want, and besides my botanical memoirs are pledged for the Lyceum.

I should like to work with you on Electricity. When will you conclude your experiments? Do post up all your discoveries and secure them in the *Phil Trans,* and then begin a fresh score. Has Faraday anticipated any of your results?

Really I am sorry that Gray cannot obtain a place in Princeton, for I *know* he would be a great acquisition. He has no superior in Botany, considering his age, and any object that he takes up he handles in a masterly manner. He will stay with me till I go to Princeton, and then I know not what arrangements to make with him. Surely [Benedict] Jaeger cannot live in Princeton on his present salary, if he was sore pinched when he had two salaries! I would not on any account drive him away though I think he is of very little use to us, but if he should take it into his head to leave us, why I should like to see a very good man in his place.

Cherries! Cherries! Cherries! You shall have some if they are to be got, I mean Mrs. Henry shall, for I would not hunt them up for you as much as I care for you.

I have several times looked for some for myself, but I could find none that pleased me, there are some to be had with the *stones* in them, but they are not so good as the others. Tomorrow, or the next day, I will make a fresh search for some.

The pneumatic trough is in my laboratory. I think it best not to have it varnished till it reaches Princeton, for it may get bruised in the transportation. You must not expect to see anything *very grand,* or you will be disappointed. It is a good article, however, and will not disgrace the college, as the old one did. I shall have it packed shortly, and take it on with me, for I must visit Princeton when our examination is over. Then I hope to make arrangements for the accommodations of my family during our residence there.

It was a good thought to hire a tinker for a servant. Pray hold on to him till I come, and then the fellow can't go if he will. He shall have his hands full; I would rather have a worker in metals than in wood. You don't say whether you will take the large shears. I can get an excellent pair, quite large, for $6!

It is time for me to stop, as it is nearly my time to examine. Present my regards to Mrs. Henry, and remind me to all our friends in Princeton

Yours very truly
John Torrey

Gray Papers, Harvard University

Maclure to Morton

March 26, 1835

Dear Sir—

Your letter of the 8 December 1834 received and I am glad to learn the prosperity of the academy and still more gratified with your intentions of having an observatory which is anathema to all our political and other rulers that such a thing has not been constructed long ago. But unfortunately all governments that have yet existed have attended to everything in proportion to their inutility and left the most useful in total neglect which requires the aid of the peoples purse which the rulers too often consider as their own and forget what they were placed in power. For that is to advance the good of the millions not to pamper and indulge their own caprices and interest. But we hope then those who produce shall have got an equal share in the distribution of all. The interests of the many will far outweigh the interest of the few and public institutions for the benefit of all be ever attended to. As soon as you have formed your plan of building and the cost approaching near the means you have of effecting it I shall give you an order on the trustees of the Germantown school for the telescope and stand and shall also subscribe something towards defraying the expense. I wrote to you sometime ago that I had some 50 to 100 boxes of minerals and geological specimens collected in the different ranges of mountains both in Europe and our union which I propose to divide between the academy, the geological society of New Haven, and the geological society of Pennsylvania. Not having an exact account of what remains many in the different shifts and changes having been lost or appropriated when I get an account from New Harmony where they at present are I shall inform of what they consist if there are more than sufficient of duplicates to fill many cabinets. They were taken in crossing the Alps and other European ranges and on the roads I travelled and lectured and with their locale. I collected them with the idea that I could make a sketch of the geology of Europe as I afterwards did of the United States and had begun a geological map of Spain which I lost. And finding much confusion even in the four neptunian formations from the different ranges of mountains intersecting one another not like ours which is all attached to one chain and exceedingly regularly apparently in the state they were originally deposited without any derangement by volcanic eruption I dropt it as too difficult an undertaking for one individual.

I some time ago had an intention of establishing a printing press in Philadelphia to reprint the other penny magazines and other cheap publications from Britain from the stereotype casts which I should have imported to give our working classes much useful knowledge even cheaper than they could in Britain. And for that purpose wrote to Mr. Alexander Grews whom I supposed took a pleasure in everything that could advance the dissemination of useful knowledge from his former habits. But not receiving any answer to my frequent letters I have for the present dropt the idea and leave it to individual cupidity having since learned that the trade have taken it up as a profitable speculation and will no doubt push it as far and further than I could possibly do. And my sole intention being to have cheap books for the millions universally circulated so as it is done I do not care who does it. For the last half age I have ceased all gratification from the physical appetites. All my pleasures passtimes and amusements are moral which I long endeavored in following. My only interests in this moral rule that as a sociable being it is my interest to have as many friends and as few enemies as possible to obtain which I must do as much good as little harm as possible. The principle has wrought well for my own self satisfaction and gratification which I have been convinced is all my egotism aims at. Fully persuaded that all my own actions are selfish it was easy to transfer the motives to that of all others and the pride of self is the centre from which all my opinions radiate as you may see if Juda Dobson has sent the academy the volume already published as he has been desired. Experiences taught me in all countries that I had access to observe and analyze that the great inequality of property, knowledge, and power introduced by civilization as far as it has yet gone is the cause and origin of most of the evils, troubles, and miseries, which torment humanity. And that the equalization of those three essentials of freedom and the shortest road to reform the vices, crimes and misfortunes, of the millions and that the equalization of knowledge was the only part that an individual could prudently attempt. And that by raising the mass of industrious producers by the diffusion of useful knowledge the only mode of accomplishing the equality on which the freedom and happiness of our species depends. At first I thought the education of children the best and most radical way of reforming men. But finding such opposition from the obstinacy of schoolmasters the ignorance of parents, and the intolerant bigotry of priestcraft it became anything but amusement to wrangle with their prejudices. And foreseeing that our working classes would adopt a radical reform in the instruction of their children which is all I wish for I have dropped all idea of prosecuting that source of amusement in our union when the three hindrances are so deeply rooted by time, pride, and presumption as to afford me pleasure in endeavouring to counteract them. And my endeavours shall in future be limited by the extending of the influence of the cheap printing press.

The want of an observatory and the existence of 705 monopolizing bank charters with the innumerable incorporate monopolies is sufficient proof of the abuses of power and property and the necessity of a radical reform. The most astonishing circumstance in a free country by universal suffrage is that of all the presidents, governors legislators, etc. Jefferson was the only [one] who had any knowledge of the exact sciences and under his presidencies the only attempt to establish an observatory was made by sending to Europe a man of science[12] to choose a complete assortment of astronomical instruments of the first crafts to be found. Which instruments are I suppose resting in some of the dark corners of our mysterious diplomacy while the wordy lawyers are occupying the floor of our legislatures with those 3 day declamations on the best way of complicating the political machine and filling their pockets with bank and other corporation fees. I am delighted and live only in the bright prospects of futurity when scarce a mechanics apprentice of the rising generation but has more useful knowledge of the arts and sciences than

[12] Ferdinand Rudolph Hassler (1770–1843) was founder of the Coast Survey. Hassler was a Swiss scientist who trained many Americans. He started the research on weights and measures which ultimately developed into the present National Bureau of Standards.

all the lawyers whose whole stock of knowledge is in the dextrous use and abuse of words in one of the great anomalies of our age and will scarcely be creditied [one word unreadable] even in this half barbarian country only emerging from a 300 year bondage to a foreign despotism though their rulers have, like all others, a great antipathy to the diffusion of knowledge. Yet an individual bookseller M. Galvan has erected an observatory on the top of his house and imported from the first artists of Europe 5 or 6000 dollars of astronomical instruments with which a certain number of the inhabitants are capable of receiving pleasure and amusement. But unfortunately the whole of this town is built on a low or loam alluvial which cannot give a solid foundation for a transit instrument and defeats many of the useful purposes of an observatory. In the erection of your tower you must attend so that you have the foundation of your walls on a solid base that the carriages or anything rolling heavily along the streets may shake or derange the perpendicularity. But I suppose you will have the best advice for the construction of your observatory. Perhaps Professor Milington may interest himself if he is not too much occupied with the gold mines of Virginia whose utility is in my opinion rather problematical as it will only augment the worship of the golden calf already carried to a most lamentable excess over our whole universe. I am fixed here I believe for life in a climate that suits me and I think most old men who have learned physical moderation and refraining from the excess of the physical appetites.

I am likewise much amused with the general progress of the moral from a state of slavery, and half barbarism to a state of the highest civilization yet attained by the exact copies they take of their neighbors. And the intercourse they have with other nations on the summit of civilization every act or science they adopt is from the most perfect model that is to be found in any country. And they are like all emerging from the half savage state accurate imitators like the Hindus and Chinies. All their properties affords mental food for an idealist who delights in the fair prospect of futurity. And from the habit of observing the follies irrational and vicious propensities of humanity from an unliked education and the corrupt practices of church and state their willingness to be instructed and absence of that pride presumption of their own perfection a great stumbling block on the way of improvement would extend the field of my only amusement and afford a more rapid circulation of improvement than in any state of society I have yet been placed in. The immense weight of the church property operating through a standing army and the remains of an aristocracy left by those old European tyrants are the only obstructions to that great advance on the scale of civilization. And as the property of the priesthood is the sole foundation of this league and combination against humanity the probability is that the necessities of the state may soon force them to remove that ban to the freedom comfort and happiness of our species when I shall have full scope to the exercise of a past time that long habit has made a pleasure. Excuse all this egotism it being the only thing that doesn't interest others and of course the subject most probably containing something new is the best thing I could think of to fill a blank page in a letter which will go by young [name unclear] to New Orleans and put into the packets save postage and not cost more than it is worth. I remain yours sincerely

Wm. Maclure

Maclure to Morton

Mexico August 1835

Dear Sir

Your letter with the late Mr. De Schweinitz memoir of the 8 June received. It was prudent to give up for the present the building an observatory until you see how the municipality and Philosophical Society will execute their intentions and perhaps it may be as well to defer any kind of building on the lot until you see how the bank mania and rage for hazardous speculations will produce, as I rather think the results will be money dearer and labor and everything else cheaper. I thank you for making me acquainted with Dr. M Burrough and hope it is the beginning of the rulers of a free people whose freedom and independence must [rest] upon exact ratio of their useful knowledge to pay some little attention to the useful and exact sciences in the appointment of their public officers. More particularly their foreign agents whose influence respectability [gap in manuscript] must always be in proportion to their general and useful knowledge like Franklin or Jefferson not from their familiarity with the local uses of law or politics, following Aaron Bur[r] through Sweden all those he had conversed with expressed their astonishment that so ignorant a man could acquire so much reputation and power amongst a well informed free people. There is little or no science in our legislators or indeed in any of the law or political rulers in any part of the world. One of the greatest drawbacks to civilization and the principal cause of the rapid advance of France during and since their first revolution the great number of men of science elevated by democracy into power.

I thank you and the members of the society for indulging my hobby in distributing my opinions to the different libraries clubs and meetings of the working classes. I have been long convinced that they cannot agree with the different ranks or be agreeable or useful to any of them. But being the result of long cogitation without which life is an unsupportable burthen I have been induced to expose them hoping some one may either contradict or confirm them. My old acquaintance Julien editor of the revue encyclopedie of Paris has reviewed them in one of the numbers of last year or the beginning of this which I wish to see. The monthly of February 1834 has taken notice of them, no [one] were inclined to perpetuate, the antient systems of education and finding too laborious unthankful and vain endeavor to push both ignorant parents and obstinate into anything rational. I have in part given up my plans of schools for which I collected my library and will assist the library of the Academy with all the books which it contains of which already they have not a copy which perhaps may principally be those on antiquities for the natural history. The most of the same are already in the academy's library. It will be necessary to mail a catalogue of those already in the library to my brother Alexander Maclure that he may avoid sending you any duplicates of those already within the academy's library and you have any members at New Orleans who will take the trouble of forwarding the books and minerals. Give this address to my brother for as yet we have found none punctual. Say's as well as our packages have lain for years without being forwarded. Bonaparte's great work on Egypt and Pyranisi [i.e. Piranesi] on the antiquities of Italy which will make beginning and not trust all to the same steam boat or ship. Mrs. Say of New York has directions to subscribe for me to all the cheap periodicals published for the use of the working classes. And after reading them to make a paper parcel every month and send it by packet if possible under the care of some passenger to Messrs Cotton Burrough and Machee of Vera Cruz who after reading that they plese of them will forward them to me. Juda Dobson charged me in his account for 1 years subscription of Littels Museum which I have received til April 8. More numbers are required to make up the year. Be so good as to inform Mr. Dobson that by forwarding them to Mrs. Say No. 170(?) Broadway New York to be enclosed in the packet she forwards I will have much better chance of receiving them than by putting aboard of any of the traders. The same for any thing you or any of my friends may have to forward. The travels of the Prince Maximilian de Niewed is here in German who was a great friend of our much lamented Mr. Say and speaks much about him. If translated as I suppose it will be I wish you to inform Mr. Dobson to send a copy to Mrs. Say along with the museum by Littel to be forwarded to me and much oblige. Yours truly

Wm. Maclure

Torrey to Morton

New York January 24th 1838

Dear Sir.

Last evening in a conversation with Dr. Pickering[13] he informed me that you have had some correspondence with Mr. Maclure, respecting a new edition of Michaux's Sylva. If you will look at my introduction to poor Croom's Catalogue of Newbern plants (a copy of which I sent to the Academy) you will see that my late friend had entertained a similar plan and he engaged me to address a letter to Mr. Maclure to ascertain whether he would aid the enterprise and state on what terms he would grant the use of the copper plates. The death of Mr. Croom prevented my writing, and it is a singular coincidence that Mr. Maclure was at the same time making arrangements to accomplish what is so greatly needed. It is very important that the work should be well done. There are many errors in Michaux, not only in the descriptions and geography of the species but in the statistics. A very considerable number of species have been omitted and these should be figured and described. That remarkable tree the Maclura was entirely unknown (botanically) when Michaux wrote, and must by all means be introduced. As Pickering is going on the expedition to the South Seas and Nuttall has his hands full I am fearful that you will not find a botanist in Philadelphia who will be able to post up the existing knowledge on N. American forest trees. If Dr. Gray should resign his situation in the Exploring Expedition (which is not improbable)[14] he

[13] Charles Pickering (1805–78), a naturalist with the Wilkes Expedition.

[14] Gray did resign and became professor of botany at the University of Michigan. Gray never actually taught at Michigan. He was sent overseas for a year to procure books in 1838–39. Because of financial difficulties the regents of the University asked Gray to agree to a temporary suspension of his salary which eventually became permanent. In 1842 Gray received a professorship at Harvard where he remained until his death in 1888.

would be just the man to act as editor of the work, and I wish you would urge Mr. Maclure to engage his services. The book would sell well, especially if edited by a botanist of reputation. Perhaps, however, Mr. Nuttall may undertake this task after he has described his western plants. It would be necessary for the editor to visit some of the great timber Districts, dock yards etc. and to travel in some of the western states. I have a large stock of materials that would be of service to the editor who ever he may be.

A few days since I received from Mr. Nicol of Edinburgh a letter relating to fossil woods. He has examined the structure of the fossil tree of Illinois described by Hennepin in 1661. Hitherto it has been regarded as *Black Walnut,* but Mr. Nicol has ascertained that it belongs to the Arancarian group of the Coniferae. You know probably that it was Mr. N. and not Witham, who discovered the method of revealing the structure of fossil woods of this country, and he has only two specimens of them in his extensive cabinet! Knowing that you had good opportunities of obtaining such specimens, and that you are deeply interested in every branch of geological science, I apply to you to aid me in supplying Mr. Nicol with specimens. Of the rarer kinds mere fragments would suffice for determining the structure. Please answer this letter at your earliest convenience, and believe me

Dear Sir
Yours respectfully
John Torrey

Morton Papers, American Philosophical Society

THE GEOPHYSICAL TRADITION I: THE ALLIANCE OF PHYSICS AND GEOGRAPHY

When the physical sciences between the times of young Bowditch and the young physicists J. Willard Gibbs, Henry Augustus Rowland, and A. A. Michelson (let us say roughly between 1810 and 1870) are surveyed, the findings are somewhat odd and disappointing. Instead of continuing activities and interrelations between the sciences, what is apparently disclosed are scattered, discontinuous researches—some of great promise and achievement, but most of them modest in terms of results. Only in astronomy is there a continuing research program of high standards.

Joseph Henry, an excellent experimental physicist, did not, however, found a school of American physicists. Benjamin Peirce, the Harvard mathematician, trained many young men but was not the source of a tradition of research in pure mathematics. Unlike these men the chemist Wolcott Gibbs of Harvard, younger than Henry and Peirce, lived into the day of the graduate school; and perhaps as a result of this he produced a line of research chemists. But the impression remains that physical scientists whose productive years were before 1870 were rather odd, isolated figures without significant disciples.

But this impression is a result of looking at the past from our present frame of reference. If we examine science in America in the last century on its own terms, a very interesting pattern is revealed in which the sciences are arrayed quite differently from the present order. Bowditch and astronomy are perhaps the best starting points for delineating the pattern which involved all of the sciences, not only the work in natural philosophy.

Bowditch was not unique in his interest in mathematics in America, although he was the most eminent of those interested. Others had training and skills in mathematics. Like Bowditch, they were generally not concerned with pure mathematics nor with the use of mathematics in the development of physical theory. Mathematics was a computational tool for applications of theory, primarily in engineering and in navigation. Eventually all of engineering practice would become mathematical, but in the first half of the nineteenth century in America most of the uses were in conjunction with civil engineering or surveying for a purpose identical with the mathematical interests of navigators—the accurate determination of the location of particular points on the surface of the earth, directly or indirectly based on astronomic data.

Astronomy always had high prestige among the sciences. It was the model by which all self-respecting scientists of that day judged the state of their fields. In America astronomy also had high repute; national pride as well as love of science made the establishment of observatories almost an act of faith among the enlightened public in the middle decades of the last century. Funds might be scarce for laboratories, but a generous Congress by the 1840's would send astronomical parties halfway around the world. This support for astronomy derived partly from its exalted status as the most exact of the sciences and partly from the belief that much of its results were useful in navigation, time determinations, surveying, and the like. A good case might be made for the view that support for astronomy stemmed as much from a belief in its usefulness as from its prestige as the pace-setter of the sciences.

But it was geography, not astronomy, which really dominated the sciences in America in the first six or seven decades of the nineteenth century. For this brief period geography was the queen of the sciences and astronomy the handmaiden to geography. Most of the sciences served similarly. We can define geographical science of that era as composed of two great divisions, natural history and geophysics. The first can be simply defined as the nonexperimental parts of the life sciences and geology, especially where the emphasis was on taxonomy. Geophysics is the study of the physical properties of the earth, including its dimensions, surface features, oceans, inner structure, and the surrounding atmosphere of the planet. Gravity and terrestrial magnetism, as well as meteorology, are in its purview. Natural history and geophysics came together in the service of geography most notably in the surveys and exploring expeditions on which, as a rule, both specimens and physical data were gathered and then analyzed. The principal difference between the two was fundamental: geophysics used mathematics both for computation and often for the language of theory. Natural history was usually not quantitative and its theories were almost never couched in mathematical symbols.

Under the banner of geography there was a kind of unity of the sciences whereby most scientists were enlisted in the eminently practical task of describing the physical and natural characteristics of the nation, as well as other portions of the earth. While the practical aspects certainly helped loosen purse strings, the scientists involved almost invariably took their researches beyond the level of immediate usefulness to the level of possible contribution to the growth of the theoretical structure of their sciences—much to the dismay and indignation of legislators on several occasions.

Most physical scientists were not, in this period, solely concerned with pure physics, chemistry, mathematics, or astronomy. Chemists analyzed rocks and minerals gathered in the field; physicists accompanied surveying parties and studied the effects of the earth's magnetism; mathematicians calculated the shape of the earth from geodetic data; astronomers studied the movements of the clouds and winds. Indeed, few American physical scientists of any consequence during this period were simply laboratory experimentalists or theoreticians within the limits of their fields. Most were to some extent concerned with the physics of the land, the oceans, the atmosphere. Even in astronomy, in which so much work took place in the middle decades of the century, the studies of the heavens in the days of the classical astronomy of position, mass, and orbits was another kind of mapping operation. Astronomers studied the earth as another celestial body. And the success of the mapping of the heavens surely encouraged the students of the earth to gather data and seek regularities which would yield laws analogous to the achievements of Newton.

What pure sciences could grow best in this environment? Four fields in the period before the Civil War were most benefited by American activity; astronomy, taxonomic botany, geology (including paleontology), and meteorology. All were part of geography. The first two were well-developed, even classical, fields with some promise of practical benefits from research; geology and meteorology were relatively new fields with great expectations of practical benefits. In addition geology and meteorology tended to straddle the division between the physical and the biological sciences, the former with its concern with fossil remains and the latter with its interest in the effects of weather on man and crops.

Scientific fields expanded in nineteenth-century America if they were part of geography and had some expectations of practical benefits. A classical field meeting these requirements would also profit from the prestige and familiarity accruing to its activities. A new field meeting these requirements benefited from the ease with which people of little formal training could enter the discipline. Fields lagged if they were less classical than astronomy and botany and had less expectation of utility than geology and meteorology. A new field like oceanography sputtered out after a promising start because there was little expectation of immediate benefit. Research in chemistry and physics suffered from three handicaps in their development in nineteenth-century America. Both were outside of the geographical sciences; both were classical in the sense of requiring formal training, placing them beyond the reach of the amateurs; and the practical applications of pure physics and chemistry were not generally appreciated even at the end of the century.

Whatever the activities under the banner of geographical science, they usually involved established sciences or the known methods of one science in a new context. By and large the range of scientific researches was within the bounds of the established and the conventional, eminently suitable for the conquest of a continent, but usually out of the mainstream of the development of the physical sciences. Only in recent years with the International Geophysical Year and related scientific work would geophysics return to as prominent a place in the scientific scene.

While Europe was undergoing one of the great periods of growth of the physical sciences, America was mostly concerned with problems of a different order. Of course, Europeans were also interested in these problems but not to the same degree as in America. Was the cultural lag of America in this respect a serious hindrance to the development of the physical sciences in the nation? Did the subservience of the physical sciences to geography deflect promising scientists from the mainstream? Perhaps the answer is yes, but it is just as reasonable to infer that the state of American cultural and economic development made it highly improbable that many individuals would go into pure physics, chemistry, and mathematics. By providing intellectual interests and positions for the scientifically minded, the alliance between geography and physics expanded and maintained interest in scientific research.

Because of the scope of the activities involved and their nature, the alliance resulted in a wide diffusion of scientific skills and knowledges. The wide diffusion had three consequences: (1) many people with these skills and knowledges but with little interest or opportunity in pure research would ultimately turn to applied fields, to the benefit of agriculture and industry; (2) when America surged forward in the sciences in the latter decades of the last century and the early decades of this century, there would exist a fairly broad base of organizations and trained individuals; (3) since the physical sciences in the alliance relied more on careful instrumentation than on mathematical theory building, when the physical sciences began to expand in America, they would ordinarily be experimental and notable for instrumental accuracy, not for the formulation of theories.

Joseph Henry

In 1830 an issue of the *Edinburgh Journal of Science* (familiarly referred to then as "Brewster's Journal") containing an article by the Dutch scientist, Gerard Moll of Utrecht, on the development of an electromagnet with great lifting power reached Albany, New York. It was read there by a thirty-three-year-old professor of mathematics and natural philosophy at the Albany Academy (roughly the equivalent of a high school today). Joseph Henry, that young professor, was stirred by the article because he had worked for two years on the same task as Moll—"the development of great magnetic power with a small galvanic element"—and had surpassed Moll's reported accomplishments. But Moll had published first, and Henry was in danger of losing credit. A note, which appears below, was sent to Silliman on December 9, 1830, and was followed shortly by an article for the *American Journal of Science and the*

Arts. It was Henry's first major venture outside the Albany community, the true start of a great scientific career, and the initial step in a series of events greatly influencing the course of science in America.

Henry's early career, with one great exception, was very similar to that of many scientifically inclined Americans of the first half of the nineteenth century. He was from a family in modest circumstances and largely self-taught in science. Like many others he was early caught up in the excitement of scientific geography, engaging in surveys of New York State, in determinations of the latitudes and longitudes of localities, in analysis of meteorological data, and the collection of geomagnetic data. After considering medicine and engineering, the two ever present vocational possibilities, Henry elected to teach in Albany. Here, or at another school, he might have remained, as did so many with similar backgrounds—sincere, half-trained disciples of science who would inspire a few boys to choose science as their life's work.

Joseph Henry was an exception in one great respect; he was an experimental physicist, a rare bird for his time and place, and one who compared favorably with the best in Europe. Henry is best understood as a great experimental physicist who arose from and shared enthusiastically in the contemporary interest in geophysics. At Albany until 1832, at Princeton from 1832 to 1846, and finally in Washington as Secretary of the Smithsonian Institution from 1846 until his death in 1878, Joseph Henry performed and promoted work in the geographical sciences. His greatest contributions to science, the investigations of electricity and magnetism, occupied but a brief portion of a long, full life. But they marked Henry as almost unique, and certainly as outstanding, in the America of his day.

The publication of the article in *Silliman's Journal* had an unexpected consequence. Henry was called upon by Silliman and others to construct electromagnets for use in teaching and in research. Today we take the technological descendents of Henry's devices so much for granted that we are likely to discount the significance of the construction of electromagnets of great lifting power in the 1830's. They were then the equivalents of our atom smashers, computers, and rockets—new, exciting extensions of human control of the unknown. To build them in a provincial capital far from the centers of industry and learning in Western Europe was no mean feat. To his American colleagues, at least, Henry in the early 1830's was a personage of consequence in science.

Although Albany had fostered his talents, Henry soon outgrew this good but minor cultural center. The state government had provided scientific employment; the Albany Institute had given Henry a sympathetic audience for his early papers, and as its librarian, he had access to a fair collection of scientific books and journals. But the Albany Academy did not provide a good environment for research. In the letter of March 28, 1832, to Silliman printed below Henry mentions the abandonment of a series of experiments (whose nature is still uncertain) when the room he used was needed for the school. When Professor John Maclean

(1800–86) of Princeton[1] approached him about accepting a position there, Henry agreed in the letter of June 28, 1832, printed below. The letter is interesting in three respects. First, Henry is quite frank in describing the disadvantages of the Albany Academy. Second, he neatly defined his sphere of interest as "intermediate to pure Mathematics . . . and to the more detailed parts of Chemistry." Third, with characteristic candor, he called attention to his lack of a college education.

Because Princeton provided a good environment for research, Henry published much in many fields. Ironically, nothing he did after 1832 equaled the achievements of his young struggling years in Albany. The move to Princeton placed him in a better position for participation in the life of the growing American scientific community. To the north was New York, not an important scientific center then but its significance would increase during his lifetime. To the south was Philadelphia, still the scientific capital of the nation in 1832. Here Henry participated in the activities of the American Philosophical Society and the Franklin Institute. In Philadelphia was his opposite number at the University of Pennsylvania, Alexander Dallas Bache, a great-grandson of Benjamin Franklin and a graduate of West Point. Well trained in science and in mathematics, Bache was interested in scientific education and in the organization and promotion of research. The two became friends and formed an alliance, as it were, for the advancement of science in America.

An almost constant theme in the Henry letters printed below and in much of the scientific correspondence of the last century is the problem of conflicting claims for priority in scientific discovery. Of course the problem still exists; professional advancement, personal pride, and even enduring fame are at stake. But the conflict is not so acute today since the scientific community has formalized the mechanisms for the determination of priority. There is now also wide recognition of the possibility of simultaneous discovery arising from the knowledge dispersed in common through the scientific community. In the nineteenth century priority determinations were less formalized nor was the ever-present threat of simultaneous discovery as generally appreciated. Priority conflicts were fairly common.

Americans were at a special disadvantage in these struggles. Distance handicapped their learning quickly of European research. If they published in America, there was always the mortifying possibility and danger that Europeans would not learn of their work or would simply ignore them. After the experience with Moll's article on the electromagnet, Silliman, and later, Bache kept urging Henry to publish both for his own reputation and for the reputation of American science. While Henry was anxious to get credit for his work and was very careful in differentiating his contributions from others, including his own collaborators, he hesitated about rushing into print and was not at all bellicose in backing priority claims.

Joseph Henry's greatest mortification over loss of priority was in the

[1] Later President of Princeton.

discovery of the induction of electricity by magnetism. The great English scientist Michael Faraday published first. For several years Faraday and Henry followed parallel courses in their work, Faraday usually getting into print first in no small measure due to the superior publication facilities in Britain. In some of the parallel experiments Henry surpassed Faraday in certain respects. Henry, on the other hand, never matched Faraday in the formulation of a theory of the relation between electricity and magnetism. In 1837 Henry went to Europe and at last met Faraday. The comments on Faraday in the excerpts from Henry's European diary and in the other letters printed in this section show Henry as a fair observer giving due credit but not overwhelmed by European reputation.

More explicit comments on the relations between American and European scientists are in the letter to Bache of August 9, 1838, which appears below. This letter is also interesting for its discussion of science in America and the proper organization of science. Popular writers and publicists might talk of the democracy of knowledge, but to Bache, Henry, and many of their contemporaries science was an aristocratic endeavor often handicapped by the egalitarian atmosphere in America.

Joseph Henry to Silliman

Albany Dec. 9th 1830

Dear Sir

I have been engaged for some time past in a series of experiments on Electro-Magnetism and in particularly in reference to the development of great magnetic power with a small galvanic element. The results I wish to publish if possible in the next No. of the Journal of science. I am anxious that they should appear as soon as possible since by delaying the publication of the principles of these experiments for nearly two years I have lately had the mortification of being anticipated in part by a paper from Prof. Moll in the last No. of Brewsters Journal

Please inform me if I shall be too late for the next No. of the Journal if I send my paper within two weeks after the date of this letter. It will probably make five or six pages. If it be not too late I should like to have a small wood cut of a powerful magnet which I am now constructing on electro-magnetic principles. . . .

Silliman to Henry

New Haven March 12, 1831

Dear Sir

I write to know whether you have anything for the ensuing No. of the Journal on galvano-magnetism; if you have made any advances you should not withhold them (in justice to yourself) as there will be other laborers in the same field. I have some short notices on the subject from Dr. J. W. Webster[2] and from Dr. Hare which will appear in April. If you have anything I should like to receive it as soon as may be.

As the river is now open I trust that our machine can come around by water and I observe that one of our New Haven sloops advertises to visit Albany soon; perhaps it may come by her but I would not wish you to hurry the affair as our medical school is now locked up but I shall have much pleasure in shewing your results to the college classes [I] remain dear sir

yours very truly
B Silliman

Prof Henry

I hope the pamphlets came safely to you and that the short abstract of your memoir in the 2nd vol. of the Chemistry is correct.

Henry to Silliman

Albany March 28th 1831

Dear Sir

I take the opportunity of my friend Dr. Powers going to New Haven to send the long promised magnet. I found it impossible with my other engagements to have it finished before the close of your medical term and therefore concluded to wait until the river opened. Dr. Powers has seen our method of operating with it and has been so obliging as not only to take charge of its conveyance but has also promised to attend the fitting of it at New Haven.

The frame we have used in our experiments is too small, we have therefore concluded not to send it. Dr. P. however can have one constructed to suit your Lecture room for but little more expense than the transportation of one from here would amount to by the steam boat.

We have also sent but one battery as the other used in our experiments (see paper) belonged to the academy. We have no time before Dr. P's departure to have another constructed.

The power of the magnet may be shown to a Class in the manner you proposed with a rope and a pulley. We have however exhibited it by piling on the scale beneath the magnet about ¾ (say 1500 lbs) of the maximum weight which it will support after showing that the

[2] John White Webster (1793–1850), a professor of chemistry at Harvard.

magnet freely sustains this by slowly withdrawing the acid from the battery we suffer the whole to fall about five or six inches. This never fails to produce a great sensation among the audience as before the fall they can scarcely believe that the magnet supports the weight. We send one large revolver the experiment would be more striking if two were used as they would turn in different directions.

The second vol. of your Chemistry was received only about six or seven days ago. I am much gratified with the analysis of my paper and am pleased that you have mentioned Dr. Ten Eyck's[3] name as you have done. In justice to myself however I must add that it is the opinion of those of my friends who are acquainted with the whole affair that my name alone should stand on the title of the paper the communication was drawn up by myself and all the experiments detailed in it except those credited to Dr. Ten Eyck were solely by me. To Dr. Ten Eyck belongs the merit of arran[g]ing the mechanical part of the apparatus.

The large magnet described in the last paper was constructed entirely by my own hands except forging the iron. The plan of the frame was made by Dr. Ten Eyck and also the drawing was made by him. The experiments with it were performed by both. In regard to the magnet we send to you, the plan was drawn by myself and the forging done under my direction, the winding with wire was done by Dr. Ten Eyck we mutually experimented with it. I have been thus explicit that you may understand what share each has had in the affair and also to answer a passage in one of your letters. I wish you would publish the account of the present magnet as an extract of a letter to you (if you consider this mode not improper). I was much gratified with your kindness in sending me fifty copies of the paper and consider myself much more than *paid for my communication* by this and other instances you have shew me of good feeling.

Did my pecuniary circumstances permit I would gladly send you the magnet free of expense but this I cannot well afford my experiments have already cost me considerable the several items of expense without counting my labor (which of course is sufficiently paid by the honor of constructing it) will amount to 35 dollars. This may if you please be transmitted to me by Dr. Powers and if you wish I will send a bill of particulars. The paper is perhaps too late for the Journal you will perhaps give it a place in the ap-

[3] Dr. Philip Ten Eyck succeeded Henry at the Albany Academy.

pendix and may find leisure to make a few experiments with it yourself before the publication of the next number.

I am Sir with much respect
Your [servant] Jos. Henry

I commenced last fall a series of observations on the intensity of magnetism at Albany and used the needles furnished to Prof. Renwick by Cap. Sabin.[4] I have since had a number of similar needles constructed and shall resume the observations next month, the results I should be pleased to communicate to the Journal with the consent of Prof. Renwick.

Henry to Silliman

Albany March 28th 1832

Dear Sir

I found it impossible to procure a proof of my paper before last Saturday afternoon the printing of the meteorological report having been deferred by a press of other matters. I fear the paper will be found too long for insertion in the present No. of the Journal. If however you cannot give it a place at this time I hope you will consider it worthy a publication in the next No. following as I wish to give it rather more circulation than it will receive in the Regents Reports. I think it desirous that the attention of observers should be now particularly directed to the subject at the present time since it is probable that the Aurora is now passing through one of its periods of maximum intensity and that the opportunity which is now offered to study its influence on the needle will soon be passed.

Please to state that the paper is from the Report of the Regents of the university in 1832 to the Legislature of the state of New York. Also make the following corrections. In the 3rd paragraph counting from the end of the paper and in the last sentence of it instead of evening read evenings. In the first foot note in the first column strike out the words *of science.*

In answer to your inquiry relative to my further experiments in magnetism I have little definite to communicate at the present time. I have not abandoned entirely the subject but have been prevented

[4] James Renwick (1790–1863), professor of natural philosophy and experimental chemistry at Columbia College, New York City. Captain, later Sir Edward, Sabine (1788–1883) was a leading investigator of terrestrial magnetism, who later headed the Royal Society and the British Association for the Advancement of Science.

by circumstances from prosecuting a series of experiments which were commenced on a very extensive plan. I had partially finished a magnet much larger than any before made and had constructed a kind of a reel on which more than a mile of copper wire was wound. I was obliged to abandon the experiments on account of the room in which my apparatus was erected being wanted for the use of the academy and it has not been convenient to resume them during the winter. I have however constructed a magnet for Prof. Cleaveland [5] of Maine of about the same power as yours and have introduced some improvements in the method of winding etc. I have also been considerably engaged in attempting to apply the principles of electro-magnetism to the separation of iron from its matrix in a manner similar to the magnetic machine now in use in the northern part of this state. My plan is very simple and I think efficient but whether it will ever supersede an improved machine with permanent magnets I am not certain. I do not feel at present disposed to hazard much in experiments on a large scale in reference to it without some definite arrangement with the present proprietor of the separating machine who has visited me several times for the purpose of conferring on improvements in his machine. I have furnished a number of Electro-magnets for toughening the magnet in the machines now at work and have I think suggested some important improvements in the construction of the separating machine. I think the apparatus could be made a very valuable affair were it properly found on scientific principles and the management of it given to a person who understood the subjects of magnetism and mechanics. It is at present in a very imperfect state although exciting considerable interest among our founders.[6]

I have given considerable attention to the subject of terrestrial magnetism and have with the assistance of Mr. Stephen Alexander[7] my relative made a number of observations on the intensity at the top and bottom of a high hill the result of this I have not yet calculated and do not know if they will produce anything interesting. The subject of terrestrial magnetism in this country affords a wide field for observation and experiment and should I be successful in procuring suitable apparatus from the Regents, I intend doing some-

[5] Parker Cleaveland (1780–1858) was teaching at Bowdoin. His was the first (1816) treatise on mineralogy and geology in America.

[6] Henry constructed magnets for the extraction of iron from pulverized ore at the Penfield Iron Works, Crown Point, N. Y.

[7] Henry's brother-in-law, who followed him to Princeton and became an astronomer.

thing myself by making a tour along the boundary line between New York and Pennsylvania to determine the change in the variation of the needle during the last 50 years. The variation for the time this line was surveyed is inscribed on stone monuments at intervals as I am informed from the Delaware river to Lake Erie.

I cannot inform you of the presen[t] strength of the acid used in producing the maximum result with your magnet. The acid and water was not measured but mixed in such quantities as to act powerfully on the zinc. Our object was to produce a saturation of magnetism and this was found to be [e]ffected in the greatest degree at the first moment of immersion of the battery to measure the power at this instant the scale below the magnet was loaded with about 1000 lbs. A heavy weight was then placed upon the end of the lever which was made to act as a steelyard with a force of 1000 lbs additional to the weight placed on the scale: the long end of the lever was lifted by two persons until the armature came in contact with the end of the magnet, the acid was suddenly raised, the lever quickly lowered and a small sliding weight rapidly moved along the lever. The point on which this weight rested before the fall of the weights was noted as indicating when added to the weights before [one word unclear] the full power of the magnet. The action of the acid did not continue more than half a minute during the experiment.

In order to produce the greatest effect the battery was not used for several hours before the experimen[t]. I should suppose that you would find no difficulty in making it support according to your plan of using it 1500 lbs. There is one position in which the armature fits best. It was not well ground and when you have it worked more perfectly let the artisan be careful that he does not bruise the wires or force them to close together so as to destroy the insulation. I had much difficulty in detecting a cause of a failure of one of my magnets which after much labour was found to be the touching of a strand of wire with another in an opposite direction.

Respectfully your obedient servant,

Joseph Henry

Henry Papers, Smithsonian Institution

Henry to John Maclean

Albany June 28th 1832

Dear Sir,

Your kind letter informing me of an expected vacancy in your College was received on the 18th instant. I would have answered it sooner, had I not been engaged in a series of experiments which I wished to announce in the next no. of the American Journal and which has occupied almost every moment of my time, not required in the duties of the Academy for the last two weeks.

To your inquiry, whether it would suit my views to accept the chair of Natural Philosophy in your institution if it were offered to me. I answer that my only views at present are to secure a comfortable support for my family and next to establish and to deserve for myself the reputation of a man of science. I have determined to confine my attention principally to a course of study and investigation intermediate to pure Mathematics on the one hand and to the more detailed parts of Chemistry on the other. Any honourable situation in which it would be a part of my duty to teach those branches and which would afford me superior advantages to those I now possess for prosecuting them will be acceptable.

I would not however readily exchange my present situation for many that might offer as it is in some respects an eligible one. The Institution is very flourishing and well established. My salary is 1000 Dollars per annum with a prospect of its being increased. As Librarian of the Albany Institute I have access to a valuable collection of scientific works and most of the European periodical publications. In connection with Dr. T. R. Beck[8] I have the principal direction of the meteorological observations made by the different academies of the State of N. Y. to the Regents of the University. In this work I am considerably interested and have hoped at some future time to deduce many facts from it of importance to the science of meteorology.

On the other hand my duties in the Academy are not well suited to my taste. I am engaged on an average seven hours in a day, one half of the time in teaching the higher classes in Mathematics, and the other half in the drudgery of instructing a class of sixty boys in the elements of Arithmetic.

If I am not mistaken in the character of your College and in

[8] Theodoric Romeyn Beck (1791–1855) was the head of the Albany Academy and one of the early supporters of Henry in his desire to have a scientific career.

the nature of the duties which will be required of me, I think I would be more pleasantly situated in Princeton than I am at present in Albany, and shall therefore accept the chair should your Trustees see fit to offer it. I could not however consistently with my feelings on the subject make much effort to obtain the appointment, but if you will be so kind as to inform me what representations are necessary to be made before the trustees will act in the affair, I will attend to them. I can refer them to Professor Silliman and to Professor Renwick for my scientific character; particularly the first named gentleman and to the Hon. Stephan Van Rensselaer and Mr. Simeon DeWitt, Surveyor General of the State of N. Y., for my private character and history. I was once a private tutor in the family of Mr. Van Rensselaer, and have enjoyed the confidence and friendship of Mr. DeWitt for several years.[9] I may also refer them to your acquaintance Prof. Green[10] of Philadelphia. If letters should be addressed to any of the above mentioned gentlemen will you have the goodness to request that they be confidential, as I am somewhat peculiarly situated, and do not wish it known at present that I have any desire to change my situation.

Are you aware of the fact that I am not a graduate of any College and that I am principally self educated? Perhaps objections may be raised on this account. I have it is true an honorary degree from Union College and have lately been elected a corresponding member of the Royal Physical Society of Edinburgh; but such honours you know are often cheaply purchased and will probably have but little weight with your trustees.

I understand of course that the whole affair is as yet only in anticipation and that even the vacancy may not take place and shall therefore make no calculation upon it until I hear something more. I have lately succeeded in a most interesting experiment that of drawing sparks of electricity from a magnet. I have some hopes of fusing platina wire by means of the same principle.

I am Sir with much respect

Yours etc.

Joseph Henry

Faculty General Mss, Princeton University

[9] Simeon DeWitt (1756–1834), a surveyor-cartographer, who served as surveyor-general of New York State for fifty years. Another early patron of Henry's, Van Rensselaer, was a large landholder in the Hudson Valley and a supporter of science.

[10] Jacob Green (1794–1841), a professor of chemistry at the Jefferson Medical College in Philadelphia. His 1827 book on electromagnetism influenced Henry.

Henry to Alexander Dallas Bache

Princeton May 18th 1835

My Dear Bache

With this I send a copy of Prof. Forbes'[11] letters which my good wife has transcribed after putting her chickens to roost. I have engaged the Corporation to make the magnetic house and hope to be ready in a few days to commence operation.

The first work will be to determine the rate of the needles sent by Prof. Forbes and next to establish approximately a meridian line; these I hope to accomplish by the aid of Mr. Alexander.

The comparison of the telescope was made on the evening after my return the night was pretty good and we had some most splendid views of the moon etc. Mr. Alexander promised to furnish a statement of the results which your committee could use at discretion; he has however gone to Albany for a few days and in the hurry of departure forgot his promise. We expect him to return tomorrow or next day when he will probably give the required statement. The reflector stood the test very well had more light but less driving power than the refractor.

You have probably received the last No. of the Annals of Philosophy and have read Mr. Faraday's abstract the facts given are almost identical with those communicated by me. I think it probable that he has given a more definite and analytical exploration of the subject than I have. He does not appeared to have used the flat spiral ribbon and consequently has not produced results as striking as those which I can now exhibit—I have devoted three or four days to experiment since my visit to Philadelphia and have succeeded in deducing some new phenomena relative to the action of the coil. When I charge my large battery with pump water or with pump water and the acid which adhered to the plates by one immersion into a strong solution of acid in order to dissolve the oxide, the coil produces loud snaps and deflagrates the metals and gives shocks for many hours and even days by only immersing in the water. This fact furnishes a very cheap method of showing many experiments in galvanism and may give the means of producing for a long time a continual stream of Electricity of considerable intensity.

Owing to the circumstance of my being disappointed in procuring

[11] James David Forbes (1809–68) of Edinburgh University, another man with whom Henry had troubles—troubles about priority—best known for his work on glaciers where he had a violent priority conflict with Louis Agassiz.

a strong magnet from Mr. Lukins I have not been able to do much in the way of magneto-electricity and shall be obliged to make a magnet for myself before going on with the subject. The magnet loaned me by Mr. L. does not hold its power.

I have read with much pleasure the description of your contrivances for shewing the radiation absorption etc. of heat and intend to direct the attention of Dr. Torrey to the paper that he may introduce this before his class. The Dr. will commence in Princeton in about a week from this time. If you wish any information relative to the *coffee* experiment I will give it in my next. Perhaps it would be well for you to make some experiments on the electricity evolved by the agitation of coffee and other similar substances. I forgot to look over your account of the facts which I communicated to the society on the evening I spent at your house as I intended; there was a mistake in one part which I do not recollect but which I wished to alter. I suppose that I can make the correction when I next attend the society.

Do not forget that you are to furnish me with a formula of observation relative to the magnetic intensity and also to order an exhausting syringe. I think that instead of a wooden bottom to the cylinder, or glass receiver, a flat plate of metal with a stop-cock soldered to the lower surface will be better. The cement can be placed on this and the whole warmed until the cement melts by a spirit lamp placed under. In this way the adjustment of the bar magnet will be less liable to be disturbed in cementing down this glass.

I will magnetize a steel bar according to promise for you as soon as I put my battery in full operation with acid.

I most sincerely hope both for your own sake and that of the advance of science that you may fully make out the disturbance of the needle on account of the eclipse. I know no fact more important or interesting in Terrestrial Physics if it can be established than the Thermo-electrical origin of the magnetism of the earth. During the next eclipse should we have to see it we must all be active in watching the needle for you.

I committed a most ridiculous blunder in writing Prof. Silliman a few days since which if you have not sent the copy of the abstract I wish you would correct. It may however have the effect of preventing the appearance of the abstract in the next No. of the journal. When I received the copy of the Annals of Philosophy and read Mr. Faraday's abstract I saw that his paper was read on the 5th of February. I immediately turned to a copy of my paper and found it dated Feby. 6th or one day later at the same time I had the impression that your abstract of my communication to the Society did not give the date of the communication but only stated "at the last meeting" and under this false impression I wrote to Prof. Silliman requesting him not to give the article into the hands of the printer until I could have time to request you to make some additions and give the date.—I have since referred to the article and much to my chagrin have found that you give the date (4th of Jan.) twice. I committed this mistake by writing in a great hurry while the mail was closing for the east.

Does not the latter part of your heading place me in rather a pugilistic position with regard to Mr. Faraday and would not it be as well in the republication in Silliman to strike out the passage *"Mr. Faraday having recently entered upon a similar train of observations, the immediate publication of the accompanying is important, that the prior claims of our fellow countryman may* not be overlook." Perhaps I am over squimish on this point and have been rendered some what more so by a remark of Prof. Green which amounted to this that there was 'much cry and little wool'. On shewing the passage to wife she says it ought to stand as it is. I will leave you to decide. I do not wish to make more of the affair than it is really worth. Please write me immediately on the receipt of this and inform me if you have sent the article to Prof. Silliman. I am very sorry to give you as much trouble as I have in reference to my affairs. You must put more duties on me in retaliation. I will send the bars of antimony and zinc by the first opportunity I find that they can be sawed and this will probably be the best method of constructing the small bars for our therm[om]et[er].

Give my respects to Mrs. Bache. Mrs. Henry will be very happy to see her in Princeton as soon as Mrs. B. can find it convenient to come on jaunt but this time could be found very pleasant. Princeton is now dressed in its most pleasing garb. The college campus is extremely beautiful at this season of the year.

I am with the highest respect
Your Friend
Joseph Henry

Silliman to Henry

Boston March 15, 1836

My dear sir

Your very kind letter of the 2 has been forwarded to me by my family and I thank you very sincerely for it: it is exactly what I wished and I now write to say that if you have any additional facts derived or soon to be derived from your lectures which I trust you are now about reforming there may be time to forward them to me before I come to that part of my course: I expect to give a lecture on electro magnetism in the last of this or the first of next month. Your drawings are excellent and enable me perfectly to understand you and if necessary I will thank you to sketch any additional experiments.

For Ampere's revolving apparatus I shall this morning send home for those which you previously furnished me with for your great magnet, although I fear they may be too many. I should like your opinion on this point taking into view my battery which is 450 pieces of 10 inches by 4, in copper cases, all plunging at once; they are in six troughs and are also fitted as a calorimeter of any number of numbers between 6 and 18. Will this be powerful enough to produce the results? I grieve to hear that you have been ill but trust in God that you are restored and I pray that your important life may be long preserved, as I think you are destined to a brilliant career in the science of this country.

It will give me the most sincere pleasure to promulgate your discoveries and I am not a little gratified that you think I am promoting a taste for science by my popular lectures.

I am now lecturing to an evening class of 1000 and to a day class, chiefly persons who cannot go out in the evening of 3 to 400 and there is great enthusiasm incited.

Hitherto every experiment has been successful. I am to go (by particular invitation) to New York on finishing here, to give a full course of geology from April 19 to May 25.

Hoping to hear from you soon at least regards your results I remain my dear sir

Yours most [one word unclear]
B. Silliman

Excerpts from Joseph Henry's European Diary

BETWEEN APRIL 3 AND APRIL 15, 1837

. . . Called at the British Museum was introduced to Mr. Gray[12] by my letter from Dr. G. of Philadelphia was kindly received. But learned something in reference to the science of our country which gave us pain. No naturalist of any reputation in the U.S. and those who propend to the science continually quarrel among themselves. Make much out of little imagine themselves entitled to immortal honor for having described a new species when things of this kind are of every day occurrance. There are now in the Museum 72 hundred species of animals not described and hence on this principle as many naturalists may be immortalized.

One naturalist (Dr. M.) came from America to see a particular animal described by Cowan[13] and brought with him some specimen in bottles and among these the very one he so much desired to behold. When told it was the animal he laughed at the assertion, gave the specimen to the Museum. Cowan afterwards himself stated it to be a genuine specimen of the animal in question.

Some of the naturalists who have visited Europe have got up a reputation out of the corps. By making communications to different societies. Thus a person with a smattering of a knowledge of shells makes a speech on the subject to the geological section is much applauded by those who know but little about the affair and in this way cheats himself and the American public into the belief that he really knows something. On the strength of this erroneous character he goes home, commences to publish and loses the little reputation he really had among those who have especially attended to the subject.

Those remarks were made with perfect freedom and although on examination may perhaps contain some truth . . .

During this week I have attended several lectures by Mr. Faraday in the morning on the metals. The subject is one of the latest in the whole course of chemistry but by his manner and the many new facts he gives the lecture is quite interesting as well as highly instructive. He is assisted by a person named Anderson who was formerly a soldier but being picked up by Mr. Faraday is rendered a very

[12] John Edward Gray (1800–75), the naturalist who greatly developed the British Museum's holdings.

[13] F. M. Cowan?

efficient person and makes a most admirable help to the active and rapid manipulations of the lecture.

Mr. Faraday in his lecture on the means of detecting the presence of arsenic insisted on the necessity of much practice before an opinion be given in the case of suspected poison by means of this metal or its oxide.

He then detailed the several processes which are given in the books but said nothing of that recently invented by Mr. March of Woolwich. I asked his opinion of it after the lecture and gathered from the answer that the method would answer as one of the proofs but that it could only be made by a person well skilled in the use of instruments and without proper precautions the whole of the metal might escape without being caught. I afterwards saw the apparatus for collecting the gas or rather generates it in the case of apparatus at Kings college and was informed by Prof. Daniel[14] that he considered their process a good one but that it is difficult to procure tin in common which does not contain in itself some arsenic but this can always be known by testing the liquid or a liquid with tin by the apparatus daily before using the suspected fluid.

In the lecture on lead Mr. F. gave a table of the different tenacities of the metals and also one of their relative hardness. To show the softness of lead an impression of a seal in wax was placed on an anvil and a piece of sheet lead of the thick kind laid on it a blow was then struck with a hammer and an impression made which would again give many new impressions of the wax. He then spoke of the alloys and the amalgams of metals and stated the fact that they are all definite compounds. When the metals are mixed in any other proportion than those proper to form the relative atomic weights a definite compound takes place among parts of the materials and the remainder form viscous real mixtures which in some cases may be separated by simple means. Thus in a compound of lead and bizmuth if not in the proper proporti[on] the excess may be melted out or rather I should say the fusible alloy will first melt and the other material which is in excess will remain in a solid state and thus separated by a little precaution in managing the heat. The silver on back of the looking glass is a definite compound and the fluid metal which runs out the quantity of metal in excess.

The fusible metal may be used for taking casts from impressed paper from the surface of any substance not easily altered by a heat

[14] John Frederic Daniell (1790–1845), an English scientist who devised the Daniell electric battery.

less than the temperature of boiling water a quantity of the metal being found over a surface of music on which the type had let a projection similar to that on the paper used for the Blind. A cast was taken which could be used as a plate for the multiplication of the copies of the music. The plates of the compound metal possess considerable hardness.

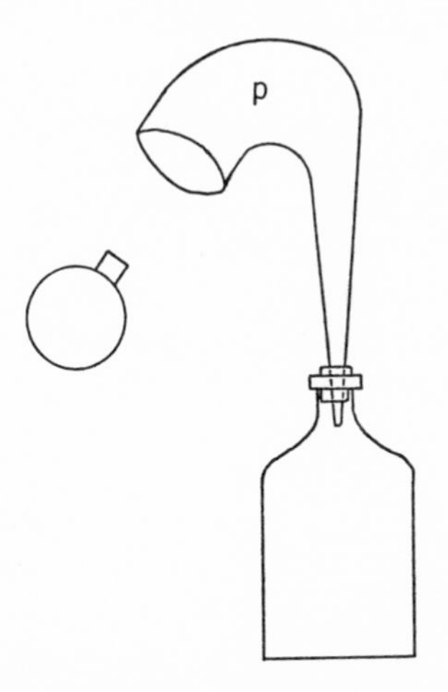

Under head of mercury it was mentioned that this metal does not expand like ice and cast iron in the act of passing into the solid state. A beautiful exp. was [performed] to illustrate the burning of mercury in chlorine. The retort was exhausted and filled with chlorine. A small quantity of fluid in [*p*] was heated by a spirit lam[p] and ignition produced.

It was stated that the combination of all metals produce heat like any other chemical union. This was shown in an extreme case by the union of potash and mercury which enter into union with heat and light.

(This general action of metals gives an explanation of the phenomenon observed when tin and platina are burned in contact JH.)

In lectures on the metals Mr. F. always puts on a blackboard behind him a table of the several compounds of the metal and constantly refers to this. The figures and letters appear to have been made by a tin marking plate. I must make inquiries of this and endeavour to procure some articles of this kind.

The precipitating glasses used by Mr. F. are of this shape [see p. 80] about six inches deep and three inches in diameter.

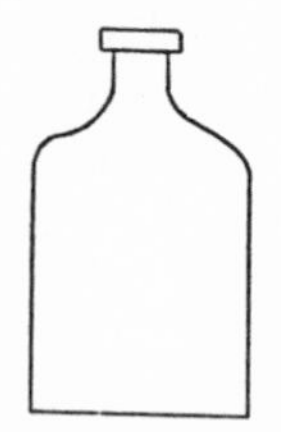

The amalgams of mercury are all very rarely decomposed and on this fact may depend the motions observed on mercury. Mr. F. spoke of the experiment of Berzelius[15] for forming something like an alloy of nitrogen hydrogen and mercury by mixing a strong solution of muriate of ammonia with [liquid] of ammonia and into this pouring a quantity of mercury. An action takes place, the mercury swells and assumes a buttery appearance. Mr. F. in this exp. made a large hole in a piece of muriate of ammonia put the mercury into this and on it poured a quantity of liquid ammonia.

The exp. was not in this way a very striking one. It has been supposed that the swelling is due to a gas given off which rising through the mercury throws the whole up in foam. I wished to ask Mr. F. if this opinion had ever been tested by making the mixture in a vacuum but did not that morning get an opportunity. Mr. Solley however informs me that he had tried the exp. thus and the result favored the supposition that the swelling was due to a frowthiness? I do not however see how this can be proved. Does it not indicate an action through the whole mass of the metal and is there not something in this peculiar? . . .

APRIL 18TH

Breakfasted this morning by appointment with Mr. Faraday was introduced to his Brother-in-law Mr. Buchanan of Edinburgh an Engineer a very intelligent pleasant man who was anxious to learn something of the rail road system in our country. I am to call on this gentleman when I visit the North. Was also introduced to Mrs. Faraday's Brother a landscape painter a Mr. Barnard.

After breakfast I remained with Mrs. F. and the two gentlemen until near eleven o'clock. Mrs. F. shewed an interesting book of pictures containing portraits of most of the Philosophers of Europe with letters from them. Also a large volume of the original papers of Sir H. Davy which were kept by Mr. F. after being copied by him

[15] J. J. Berzelius (1779–1848), a Swedish chemist who was notable for his isolation of new elements and determination of atomic weights.

for the press. These interesting reliques of departed greatness could not but inspire one with admiration. I was also shew an original sketch by the same intended for the artist to copy for one of his lectures. It did not exhibit much talent in the drawing line.

Mrs. F. is quite an intelligent and good hearted lady. She has promised to accompany me on my visit to Mrs. Sumerville.

Mr. F. showed me the room in which the great battery of Davy was kept and also the calorimeter with which the first rotation of electro-magnetism was produced. It is a very large article consisting of plates at least 2 feet by 18 inches; there are however but few of them. The room in which the great battery was kept is directly under the Theater. The wires were passed up through the ceiling. . . . Mr. F. spent some time with me this morning in shewing the arrangements of the great battery of Sir H. Davy. . . .

APRIL 24 OR 25, 1837

I was happy on these two occasions to have an opportunity of seeing Mr. F. experiment in private and was much pleased with his method. It is however altogether of the tentative or empirical kind he attempts many experiments and collects facts with great rapidity. These he afterwards arranges but does not deduce many experiments (a priori) from known principles.

The principal peculiarity in his character is his rapid and happy invention of expedients for the production of a result.

Articles of the most common are used with success to produce the most wonderful results. Wishing to [demonstrate the] decomposing power of the thermopile and having no platina poles at hand, two gold sovereigns were put in requisition. To make an extemporaneous thermopile a piece of copper wire was twisted around each end of a piece of platinum wire of about the same diameter.

Mr. F. appears to be deeply imbued with the true spirit of philosophy. I remarked that our only object in investigation should be the determination of truth, his answer was yes but this is the most difficult to learn.

Henry to Bache

Princeton Aug 9th 1838

My Dear Bache

I start tomorrow morning at ½ past 3 o'clock to put on board the Ship Toronto, the magnetic needles which, I have vibrated in this country and which I hope will reach London in proper time to meet you there, just before you leave for America. I do not know when you have concluded to sail but suppose you will start just after the meeting of the British Association next month. I send the package directed to the care of Mr. Vaughan and will request the Capt. of the Toronto to deliver it with his own hands.

I have not made out the observations but send you an exact copy of the records. The observations are made according to the old method since I was most accustomed to this. I concluded there would [be] less danger of mistake. In the series of observations made in London I was assisted by Mr. Robison the Instrument maker of Devonshire street we found no difficulty in identifying the place [of] observation as described in your letter. The observations in this country were made at Princeton with the assistance of Mr. Alexander. The results with Rusty and Bright from your observations in London and for mine and Robison at the same place you will find do not well agree, one cause of the difference is the long exposure the needles were subject to in the interval. Another and probably the principal cause is that when we came to examine the needles one of which as you may recollect was fastened with the lining of my hat and the other into the tail of my coat, the stirrups which were formed of the foil were rubbed into pieces. We collected the fragments weighed them with accuracy and substituted an equal weight of fine platina wire. The compensation was not however complete since it was impossible to give the mass precisely the same distribution as in the case of the tin foil stirrup you will find in the square box two additional needles made by Robison. He neglected to mark them and the letters put on are by myself. There is a possibility that I have not given them the proper mark, that is the one called A perhaps should on the record be called B. You will however easily determine this by a comparison of the times. I did intend sending the needles by the last London packet but was not able to get to the ship in time. I however preferred to send them by the Capt of the Toronto since I came home with him and can put more confidence in his care of the articles. They will in all probability reach London at a proper time before your departure and with the least delay. So much for the needles and now for other matters. I have twice visited Philadelphia since I wrote to you concerning my visit to your mother's. One was a short visit of a day or less. The other was with Mrs. H. and consisted of 4 or 5 days. My time however was continually occupied with some experiments in Dr. Hare's laboratory on electricity with his large electrical machine and battery. Since this was at the first of May and since then I have heard nothing from the city. My time since my return to America has been constantly occupied and since the commencement of the last college year I have had scarcely a moments leisure. Not only my ordinary duties were to be attended to but also much was expected in the way of working up the leeway of my long absence. Besides this I commenced during the spring vacation a series of experiments on electricity which have engaged my attention at every moment not devoted to professorial duties. I hope you will pardon me for not having written to you frequently as I promised and as I really intended. I have commenced since my last several letters but did not finish them at the time, did not like to send a short letter procrastinated from day to day suffered much from stings of conscience on account of the same consoled myself with the reflection that I had heard nothing from you etc. etc.

I have now again got into my old track and things go on with me something after their usual fashion. When I first returned from England however I was much dissatisfied with many things both in reference to my own affairs and those of the public generally. An absence from a place for a single year makes many changes and you will probably find, My Dear Bache, that the two years you have been absent has made as great a change in the disposition of things which you once controlled as if you had been consigned to the grave during that period. Persons with whom you have associated and perhaps controlled have learned to do without you and measures and plans which you have laboured to advance are set aside for others not as good but carry into operation because the production of others and those on the spot. Besides this the charlatanism of our country struck me much more disagreeably when I first returned than before or even now. I often thought of the remark you were in the habit of making that we must put down quackery or quackery will put down science. You have probably heard of the wonderful sensation produced in the country by magnetic machines a company was formed in New York which succeeded in [procuring] $12,500 dollars for experiments on the machine and after much puffing and the expenditure of the above mentioned sum the whole fell through. Again a great sensation was produced by the Magnetic Telegraph of Professor Morse[16] who first claimed or his Friends for him the entire origin of the project. But the most disgusting form of charlatanism which has

[16] Henry had already demonstrated a telegraph and had given Morse crucial advice.

been got up in this country since your departure is that of Dr. Sherwood in connection with the committee of naval affairs in the senate of the United States. Dr. S. brought before congress last session a great discovery in magnetism no less than that of the solution of the whole problem of terrestrial magnetism a ridiculous and puerile affair. It was however referred to the committee of naval affairs who reported on it in the most flattering terms, stated that from the opinion of several scientific gentlemen as well as their own examination the discoveries and inventions of Dr. Sherwood were of the highest importance, worthy the confidence of the public and the patronage of congress. They proposed to bring in a bill for the reward of Dr. S. but fortunately for the honor of the country congress adjourned previous to this disgrace being inflicted on the country. I will attach to the package a copy of the report for your inspection. 5000 copies extra of the report were ordered printed for the edification of the people of the united states and of the world, for I have no doubt but many of these will find their way across the Atlantic. This article you will say is a disgrace to the country. I have given a notice of it and made a protest, in behalf of the scientific character of the united states, [ag]ainst the custom of publishing scientific articles among the documents of congress before their true character is ascertained. Bad as this is it is not quite equal to a memorial published among the documents early last winter on the subject of the explosion of steam boilers. The author set forth the wonderful fact that the explosions were generally produced by the generation of *negative* and positive electricity in the boiler!! I think you will be somewhat displeased with several of the selections for the Franklin Journal since your departure. Silliman's Journal in the last two numbers is tolerably decent but contains little that is new or interesting but for the year before this it was filled with a mass of trash relative to electricity and electro magnetism which would disgrace the annals of electricity. One Fellow actually stated in a paper on electro magnetism that he had succeeded in making the magnets by substituting copper instead of iron within the coil. I had an interview with Professor Silliman and complained to him of the injury done to the character of American science by such publications. He said if I could see what he rejected I would scarcely complain of what he inserted. I also hinted the importance of having colaborators in the different departments of science for a journal of this kind. The hint was not however taken and shortly after the name of the Professors son was attached to the Journal. I have the highest possible regard for Professor

Silliman and feel much attached to him for his kindness to me and his readiness to give me often much more than my proper share of credit. Still I am now more than ever of your opinion that the real working men in the way of science in this country should make common cause and endeavour by every proper means unitedly to raise their own scientific character. To make science more respected at home to increase the facilities of scientific investigations and the inducements to scientific labours. There is the disposition on the part of our government to advance the cause if this were properly directed. At present however Charlatanism is much more likely to meet with attention and reward than true unpretending merit.

But I fear I will tire you with this long story of ills and proposed remedies and I will now change the subject to that of the British association which you are about to attend. You have seen the report of the last meeting in the Atheneum and found honorable mention of your own name there. My recollections of the meeting on the whole are pleasant. There were however some circumstances which were at the time not very agreeable and which I mentioned briefly in my last letter. The principal of them was a slight altercation I had with Dr. Lardner[17] relative to the speed of American Steam boats. I had resolved to keep myself perfectly cool and not to put myself too much in advance. I concluded, on account of my character at home more than from a desire to show myself at the meeting, to give the mechanical section a few words on the subject of the internal improvements in the United States and to the physical section an account of my researches on the lateral discharge. I had been requested by Mr. Tanner[18] of Philadelphia to present a map to the society containing all the lines of railways and canals in the country up to 1836. In presenting this map I occupied the time of the section about 6 or 7 minutes on some general views relative to the topography of the country. The extent of navigable rivers, canals and railways constructed etc. My remarks were listened to with much attention and received with great applause; but just as I was stepping from the platform a person among the audience arose and requested to ask me one question in reference to the speed of American steam boats. He said he wished authentic information. The chairman said the question was not strictly proper and I might answer it or not as I chose it did not belong to the subject of my communication.

[17] Dionysius Lardner (1793–1859), a British polymath and popular writer who in 1837 expressed doubts about whether steamships could cross the Atlantic.
[18] Henry S. Tanner (1796–1858), a well-known cartographer.

I stated that I had no objection to give what information I possessed on the subject, that I was not a practical engineer, although I had paid some little attention to steam navigation in the United States, that I had lived on the banks of the Hudson and had been in the habit of going up that river in the swift boats and that I had gone the distance of 150 miles from N. York to Albany up the river in the space of 9 hours but that there was a tide in the river from perhaps 2 to 3 miles an hour at the time which assisted the passage. Dr. Lardner then arose and stated that the section did not want popular information of the kind I had given and that he did not believe that a vessel was passed through the water with the velocity I had stated. The chairman Dr. Robinson the astronomer interfered and made some very severe remarks on the turn the affair had taken, the impropriety of treating a Foreigner who had favored the association with a communication in such a manner, etc. etc. Fortunately I did not get in the least degree excited although the whole room was in a state of commotion. I next made some explanatory remarks and stated that in reference to science I did not consider myself a foreigner, that truth and science was of no country and that I wished no more courtesy shown my communications than those made by the other members of the association. I next made some statements relative to the length breadth etc. of our boats, quantity of fuel burned, etc. I then left the room when Mr. Russel of Edinburgh the wave man took up the cause and informed the section what I had done for science, etc. etc. The affair though very unpleasant at the time and entirely unprovoked and unlooked for by me, ended rather pleasantly than otherwise since during my stay in Liverpool I was constantly with marked attention and many persons introduced themselves to me and expressed their regret that I should have any cause to think that America and Americans were not highly esteemed in England and particularly in Liverpool. When my paper on electricity was presented at the Physical section the room was much crowded and the communication was received apparently with much interest.

My opinion however is that it is not in very good taste for a stranger to occupy much of the time of the section and had I not been urged to the affair and supported in it by the example of De La Rive[19] and Justus Leibig,[20] the chemist, I would have hesitated

[19] Auguste Arthur de la Rive (1801–73), a Swiss scientist also working in electricity.

[20] Justus Leibig (1803–73), a great pioneer of organic chemistry.

to offer anything. I however took good care to make my communication as short as possible, my drawings on the black board previously prepared and my lesson well covered. I was much pleased at the association to make the acquaintance of Professor Moll, of Utrecht. He promised to write me and send a copy of an engraved portrait of himself but I have not yet secured the letter or the article. Daniel[l] and Faraday were there and on every occasion treated me with marked attention. Faraday was of great importance to me since he made it a point to come to me even through a crowd and introduce me generally to all around. Then introductions I found had more weight than even Mr. F. appeared to attach to them. I found it necessary to walk with much circumspection and can easily see how our Friend Dr. H.[21] of Philadelphia erred in his intercourse with the science of Great Britain. There is a great prejudice and perhaps in some respects a just one (from the persons who have visited England) against Americans. They treat us with great kindness and I have no doubt but many of the persons with whom you and myself associated had a real respect for us and were prompted by proper feelings to show us the attention we received but still there was in my case the appearance of a little hesitation in allowing the same merit in public as in private. You may probably remark something of the same kind. It may arise from the influence of the aristocratical and political Institutions of the country and the general low opinion which is maintained relative to American science and literature.

I think you will not be much impressed with the scientific character of the British association there is such a mixture of display of ignorance and wisdom, of management in the compliments given and the honours received that the whole makes rather an unfavorable impression on a person admitted a little behind the scenes. I was at first a little at a loss to know what good was promoted by the meeting. A little reflection however showed me that the principal advantage of the Institution is the amount of money it gives to real working men for the prosecution of their respective branches. Thus money in the hands of the committees and they are wisely composed only of those who have some reputation for science. The great body of the members have no voice in the management of the Institution and in this respect the society is quite as aristocratical as the government of the nation.

This arrangement however I am far from considering improper on the contrary were it otherwise the *third* and *fourth* rate men

[21] Robert Hare.

would soon control the affair and render the whole abortive and ridiculous. Much has been said since my return of the propriety of a meeting of the kind among us, but I am convinced a promiscuous assembly of those who call themselves men of science in this country would only end in our disgrace. At the closing of the meeting votes of thanks were given in some form or other to almost all the principal men connected with the association. The names of four Foreigners were mentioned namely Moll, De La Rive, Leibig and my own. Also a vote of thanks was given to our minister Mr. Stevenson[22] who rendered himself very popular by a speech which he made at one of the dinners. I was pleased at this since when he first arrived I feared he would have made himself ridiculous with some remarks he was about to make at the Physical section on the subject of Light. I told him rather plainly that the subject did not belong to his line and and that he knew nothing about it. At this he appeared somewhat offended. His speech however made proper amends; he left the association well pleased with himself and with the good opinion of all who heard him.

I informed you that I had visited Sir David Brewster, staid two days with him and was highly delighted with my visit. He gave me a letter to Dr. Sprague[23] of Albany which contained some remarks relative to American Science and make honorable mention of our names. Dr. Sprague proposed to me to publish the letter but as I did not know how you would be pleased with the notion I thought it not worth while without some special reason to take this step. I found your Friend Professor Stevely [?] a very fine fellow and I am much indebted to him for his kind attention to me while at the association. We will however be able to form a more proper estimate of the different persons when we come to compare notes.

But to turn to another subject, do you recollect the project we had of procuring for this country a set of the Elgin marbles? On my return [from] England I spoke to Mr. [Childress] on the subject and he gave directions how to proceed in the affair. To get some of the most influential persons of Philadelphia to join in an application and if possible to have this backed by the government. On my First visit to Philadelphia after my return I called with Mr. Rogers on Mr. Biddle, informed him that while we were together in London we had

[22] Andrew Stevenson (1784–1857), who became rector of the University of Virginia on his return from London.

[23] Possibly William Buell Sprague (1795–1876), a clergyman best known as a collector of autographs.

been informed that the marbles could be obtained without much difficulty by a proper application etc. He appeared pleased with the affair and said he would give the matter proper attention. He however appeared to misunderstand the proposition for since it came from you he concluded that the copies were to be obtained for the Gerard [Girard] College and said that he would place the matter before the board of trustees at the next meeting. I again called on him relative to the same affair but he was then much engaged and said that the gentleman sent to this country by the Bank of England had promised to attend to the Subject on his return to England and that he had seen a short time before an account that any person could procure copies by merely paying for the expense of the work period. Here the affair ended and I have had nothing more in reference to it. Will it not be well for you to make some further inquiries relative to the expense of procuring; the mode etc. in order that something may yet be done to procure a set for this country.

While I was in London my friend Henry James[24] wished to procure something as a present to the Albany Institute. I advised him to get a copy of the Rosetta Stone from the museum and gave him a letter to Mr. Gray with the request that he would assist my Friend to procure the article. The stone was copied the whole cost was only 5 dollars and the Institute was delighted with the present and the papers from Georgia to Maine proclaimed the glorious deed.

Our new house is just finished and we are about to enter it or at least hope to move in a few days. The college is still in a flourishing state, the University of Pennsylvania is also in tolerably good condition. Dr. Ludlow I think is increasing in popularity and gradually extending his influence. Your successor appears to be a very kind and worthy man but exceedingly queer in his manner. He is said not to be very successful as an experimenter in the way of Chemistry and natural Philosophy before his class. Professor [Henry] Vethak[e][25] has lately published a work on political economy. He comes out against the Banking system of the country and in favour of protecting literary as well as other property. I have heard nothing from Dr. Lock[26] since my return except a communication of his in Silliman

[24] The father of William James, the psychologist, and Henry James, the novelist. Joseph Henry had tutored him in Albany.

[25] Vethake (1792–1866), an economist was at the time of this letter professor of mathematics and philosophy at the University of Pennsylvania.

[26] John Locke (1792–1856), a physician who did research in geology and paleontology as well as electricity and magnetism. He also invented scientific apparatus.

and another to the American Philosophical Society on the dip variation etc. etc. [Joseph] Saxton[27] has a good situation in the mint. He spent a few days with me in the Spring at Princeton, we had a very pleasant time in Experimenting. Professor Daubeny[28] started for Europe on the 8th of July. I did not have the pleasure of meeting with him while he was in the country; Dr. Torrey is much engaged in preparing a general Flora of the United States, one volume of the work is now in press. The scientific expedition it is said will sail in the course of a few weeks. It is now under the command of Lieut. Wilkes, Johnson, Coat[e]s and several others have been discharged in the process of cutting down the number of scientific corps.[29] You have probably heard by the papers that this summer has been one of the warmest for many years, the thermometer at Princeton has stood at 100° in shade several times. Pexii has treated me in a very shameful manner. I purchased of him a large collection of articles which I paid for in advance, these he promised to send to New York so that they would reach the Custom House about the time I might be expected to arrive. I waited very impatiently for them all winter and have only received them within about 6 weeks of the present time. When I came to examine the invoice I find that he has still not forwarded a part of the order and I know not when he will send the remainder. Also when the chemicals bought of Robequet were unpacked there was a deficiency of articles to the amount of 200 Francs. So much for the honesty of the French.[30]

[30] End of text in the manuscript.

Henry Papers, Smithsonian Institution

Torrey to Henry

New York Novr. 9th 1838

My dear friend.

My assistant Mr. Bourne, (who is an excellent fellow and will do well) brought the mirror safely but the cage is missing, nor is there any ring for holding a ball. He lost nothing so that if there was any

[27] A notable American designer of precision instruments (1799–1873).

[28] Charles G. B. Daubeny (1795–1867) was a professor of chemistry at Oxford who toured the United States in 1837, publishing an account of this trip the following year.

[29] For Charles Wilkes, see the chapter on the Wilkes Expedition. Walter R. Johnson (1794–1852) had been at the Franklin Institute and the Pennsylvania Geological Survey. He would later do research for the Navy on coal. Reynell Coates (1802–66), an example of the physician-naturalist.

other article belonging to the apparatus it would have been left in Princeton. Please send it if there as soon as you can. These are chances occurring almost every day and if left at Chilton's I shall get it soon. By some blunder I lost my honest Chap Goodrich. He misunderstood my last letter and supposed I did not wish him, whereas I was greatly disappointed when he told me that he had gone off in rather a sad state of mind to the west to seek his fortune. This new one I hired, and he is of excellent family, and hungry for knowledge.

I am pleased to hear of your new discoveries in magnetism galvanism. You will doubtless let us know all about them in due time.

I rather think with the Philadelphians, that we can hardly get up yet an [appreciation] for the foundation of science. There is indeed too much charlatanism in the Country, and enough to overpower *us modest* men.

Dr. Gray did not get off till this morning. He was detained yesterday by adverse winds. He got your money will doubtless faithfully attend to your commissions. He will make a good impression whereever he meets with men who are capable of appreciating him.

If you expect me to come on in the spring, do try and have the lower room fitted up for the Class. It is a disgrace to the College that this matter is left unattended so long, and not only shows one plainly what estimate is put on my branch, but exposes to the public the little interest taken in science by the College.

The University Professors are out with their pamphlet, but I don't know whether it will do much good. All but a very small clique understand the whole business very well. It may be of some importance however, to put the history in a permanent form.

I went to see some of Davis's plans of the Michigan University a few evenings since. It will be rather an imposing gothic building, but I fear that Davis will not be able to make *working plans,* and that some grievous blunder as to the disposition of the rooms etc. will be made. I gave my views respecting the accommodations for the Chemist. I hope you will give yours when you come here to lecture. It is time that you commenced preparing those lectures. Let me urge you to *write out the lectures in full,* and have them at your fingers ends and then make an abstract of them. Let all your experiments be numbered *don't show too much.* You will do well *if you don't try too hard!*

I wish you would lend me for a few days a few articles of electromagnetic apparatus, after you have completed the notes that you commenced for me. My class has never seen the electro-magnetic spark.

I have not procured these things because they don't properly belong to my course and also because I cannot yet afford to buy them.

When you come to the city I shall of course expect you to use my house as if it were your own. With very kind regards to Mrs. Henry, I remain faithfully yours

John Torrey

Historic Letter File, Gray Herbarium

Meteorology—The American Pioneers

Meteorology in the first half of the last century is an interesting example of a field in transition from folklore to science. The weather is an inescapable fact that has always commanded interest. By 1800 this universal concern was about to engender a distinct scientific field. The study of the atmosphere and its constituents had become a major concern of science; research on heat was soon to give birth to a great branch of physics, thermodynamics, which would contribute much to meteorology; and a substantial number of astronomers and physicists were becoming interested in the problems of meteorology.

Americans played an important role in this transition from folklore to science, perhaps one of greater significance than their contributions to any other field of science in the first half of the nineteenth century. The American interest in geophysics, following in the trail of similar European investigations, encompassed the atmosphere of the Earth and how the winds and rains affected living matter on the surface of the planet. Meteorology attracted both geophysicists and natural historians in their common pursuit of geographic knowledge. An extraordinary number of individuals, scientists and nonscientists alike, kept meteorological journals and diaries with carefully recorded data on temperatures, pressures, and winds. George Washington and Thomas Jefferson kept weather records; so did the historian Abiel Holmes, the father of Oliver Wendell Holmes, Sr. It was the proper thing to do if you were an educated person and a friend of science in the late eighteenth and early nineteenth centuries.

Another reason for the popularity of meteorology was its nonacademic nature. As a new and still simple science, it was open to people with little formal training. The professional scientists had not yet established a body of knowledge and doctrine required for initiation into the guild. Although many amateurs and persons from other scientific fields flocked in, the results at first were not notable. But men of talent soon made significant advances which ultimately transformed meteorology into a highly specialized field.

The letters of three such pioneers appear here and in a subsequent section. The first, William C. Redfield (1789–1858), is a good example of the amateur scientist. He was interested in steam engineering and in the promotion of steam transportation on sea and on the land. Like so many of his contemporaries, Redfield was not confined to any one pigeonhole. He did excellent paleontological work on fossil fish. In 1848 Redfield was elected the first president of the American Association for the Advancement of Science, a most diplomatic choice since he had a foot in both the natural historians' and the geophysicists' camps. In 1833 Redfield published an analysis of the nature of hurricanes. From fallen trees and other evidences, he correctly deduced that the winds in hurricanes circle about a common center. As we shall see, he was primarily interested in the rotary motion of the winds; it was his obsession.

His great rival, who incorrectly backed elliptical wind patterns, was James P. Espy (1785–1860) of Philadelphia. In this and his other meteorological views Espy had the backing of many scientists, among them Hare and Bache, in the city that was still the scientific capital of the nation. But Espy was not primarily known for or concerned with work on the winds. Heat effects were his specialties. Where Redfield observed the results of storms at different points of the Earth, Espy applied existing physical knowledge to an analysis of the data and even performed experiments. Espy is primarily known for two things in the history of meteorology. He correctly described the cause of precipitation as due to the upward movement and consequent expansion and cooling of moist air. And as a tireless propagandist for meteorology, he promoted the gathering of meteorological data and the development of weather systems.

A younger man, Elias Loomis (1811–89), provided Espy with a great example of what a comprehensive weather system could achieve and also typified the growing sophistication of the subject. Loomis was a well-trained mathematician-astronomer-physicist from Yale, who spent a year studying in Paris. He taught at Western Reserve University (then at Hudson, Ohio) from 1837 to 1844, at New York University from 1844 to 1860, and thereafter at Yale. Loomis gathered data from many points in the eastern United States and elsewhere about "his favorite storm" of December 20, 1836, and later about two similar storms in February 1842. Applying his basic knowledge of physics and mathematics to the readings, Loomis gave a minute description of the storms and derived generalizations on the behavior of the air that read, as one recent meteorologist put it, like the work of a modern "advanced air mass analyst." Loomis also drew the first modern weather chart. His comments in the letter of October 22, 1838, to Redfield, which appears below, make it evident that informal relationships between the sciences and between the amateurs and the professionals were no longer acceptable.

Of course, no great change came about in an instant. In 1840 a young English naturalist, Charles Darwin, read with interest Redfield on whirlwinds and opened a correspondence with him. Moreover, like Redfield, Darwin was also interested in fossil fish. To many of his contemporaries he seemed a dilletante trying without system to bring all nature into one system. Even after his book of 1859 on the origin of the species, some of his opponents accused him of a lack of systematic professionalism, by which they meant adherence to their own dogmas. By 1940 biologists and meteorologists dwelt in two different professional worlds. But in 1840 two gentlemen interested in nature and in the advancement of knowledge still shared a common scientific language and were engaged in a common intellectual venture.

William C. Redfield to Henry

New York May 31st 1837

To Professor Joseph Henry now at London
Dear Sir

I forward to you by the ship "*Mediator*" Capt Champlin a copy of "Hall's statistics of the West," which I have at length succeeded in procuring from Cincinnati, but which I apprehend will not quite meet your expectations.

I also forward to you a few more copies of my reply to Mr Espy, together with several copies of some short sketches in Meteorology an abstract of my former observations on Storms etc. with a short article on currents, in the form in which they appear in the forthcoming edition of the American Coast Pilot. A proofcopy of the sketches I believe was also forwarded to you with my letter by the packet of the first of May. I have marked in your copy of the sketches some emendations, in view of the probability of a reprint in England, in reference to which and to my views thereof, I beg leave to refer you to a copy of a letter addressed to the Editor of the Nautical Magazine which you will find enclosed with the pamphlets. If you can interest yourself in this matter so far as to learn what course it takes I shall esteem it a favor, and if nothing of the sort will be attempted, as is most probable, I suggest the propriety of offering the pamphlet as copy for reprint in the London Atheneum, Literary Gazette or some other of their sub scientific newspapers. Capt Beaufort [31] hydrographer to the Admiralty, (who probably owns the plate of my chart which was prepared for the "Nautical") will probably lend the plate for republication, or the editor of the Nautical, in case it belongs to that periodical. A stereotyped cut for the figure on the Page I will enclose you with the pamphlet, and if

[31] Capt., later Adm. Sir, Francis Beaufort (1774–1857), hydrographer of the Royal Navy.

their map is reprinted I should like to have the whirlwind figures which are found on the American Map (on tracks I. V. and VII) transferred to the English map, to correspond with the text. I am a little apprehensive however that too much *Jonathonism*[32] will be discovered in this proposition to allow of its acceptance on that side of the water; but these misgivings may perhaps prove to be unfounded. Apropos of this matter, I wish you to notice the article on Professor Olmsteads[33] shooting stars, which appears in the Saturday Magazine for the last of November 1836, and which is published by the Society of Useful Knowledge, which is headed by Lord Brougham.

Having recently been called to make a journey to Washington, I availed myself of the opportunity of spending a few hours in Philadelphia, which were passed away pleasantly. I saw Dr. James Mease, Dr. Robert Hare, Prof. R. M. Patterson, Prof. Robley Dunglisson, Prof. Parke, Dr. Charles Pickering, Mr. John Vaughan, Dr. Emmerson, Mr. S. C. Walker[34] etc. and am much indebted to several of these gentlemen for their politeness. Prof. Parke called with me at Mr. Espy's, but we did not find him at home, and I afterwards learned that he was then lecturing before the American Lyceum. A good deal of interest appears to be awakened in Philadelphia in the subject of Meteorology, at least in the circle in which I happen to fall. Since I say you an additional article of Mr. Espy's has appeared January No. of the Franklin Journal in which our friend attempts to press two or three more storms into his service, but on referring to the facts in these cases, as I find them on record in numerous instances, and from unbiased sources, I find, that like Salmugundi's ship, those storms obstinately refused to be blown up into the sky by a concentrating wind from every point of the compass. It is not my present intention to reply publically to these allegations, as I am quite willing that Mr. Espy's exertions should earn him every degree of encouragement to which he may be thought entitled. He has been lecturing before our legislature at Albany this

[32] From Brother Jonathan, then the national symbol before displacement by Uncle Sam, thus nationalistic bragging.

[33] Denison Olmstead (1791–1859), a professor at Yale who acquired a high contemporary reputation for his work on meteoric showers.

[34] Prof. Parke is probably Roswell Park (1807–67) of the University of Pennsylvania. Emmerson is probably Gouverneur Emerson (1795–1874) of the Franklin Institute. R. M. Patterson is the son of the Patterson Bowditch corresponded with. Of this list of nine persons interested in meteorology, four (Dunglison, Emerson, Pickering, and Mease) were medical men and four (Patterson, Park, Walker, and Hare) were physical scientists. Vaughan is best described as an administrator.

spring, with a view to procure legislative aid and assistance in his inquiries.

You will doubtless have an interesting time at the Liverpool meeting of the British Association in August. In case the sketches should not be previously published in England, perhaps you may think it best to bring the substance of them to the notice of the section on Meteorology; but of the propriety of this you can best judge.

I have had the pleasure of seeing one of your fellow passengers in the Wellington, but was sorry to hear that you suffered much from sea sickness. Wishing you a full freight of European knowledge and improvements and a safe return to your country and friends, I remain your friend and obedient servant

W. C. Redfield

Redfield to Elias Loomis

New York December 2nd 1837

Dear Sir,

I have seen a copy of the Philadelphia Saturday Courier of Nov. 25th containing your notice of the late tornado at Stow [Ohio] together with comments on the same purporting to be editorial. The interest which I have taken in natural phenomena of this kind induces me to trouble you with some inquiries on a point which does not appear to be particularly noticed in your communication.

The leading fact which appears to be common to the path of nearly all such whirlwinds, viz. that the trees in all portions of the track which are exterior to its center, have usually fallen in lines of direction which, if continued, will be found to cross the center of the track, and which appears to have been so *unexpected* to our Philadelphia friends, you have fully recognized. But so far as I have noticed, you have said nothing of the comparative direction of these lines of penetration as relate to the two opposite sides of the track; leaving us, perhaps, to infer, with the Philadelphia meteorologists, that the comparative direction (as regards the line of progress) is the same on both sides of the track.

I am particularly anxious to ascertain whether such an inference can be warranted by the facts which have been observed at Stow, and have therefore ventured to draw your attention to this point, for I apprehend that it is by a comparison of this kind that we are enabled to decide whether a phenomenon of this sort was, or was not, a whirlwind; and, if a whirlwind, whether it revolved to the *left,* or *right,* as regards the position of a person standing in the center of the track.

If you are not fully confident as regards these particulars it would gratify me much if you could renew your examination, which, even at this season of the year, can doubtless be satisfactorily done, if you should find such an examination to be either necessary or convenient. I submit here a rude diagram of the tornado which visited New Brunswick, and in other cases, which may serve to show you more fully the sort of difference, if any, which may be expected to exist in the direction of objects which have fallen on the two opposite sides of the path of the whirlwind.

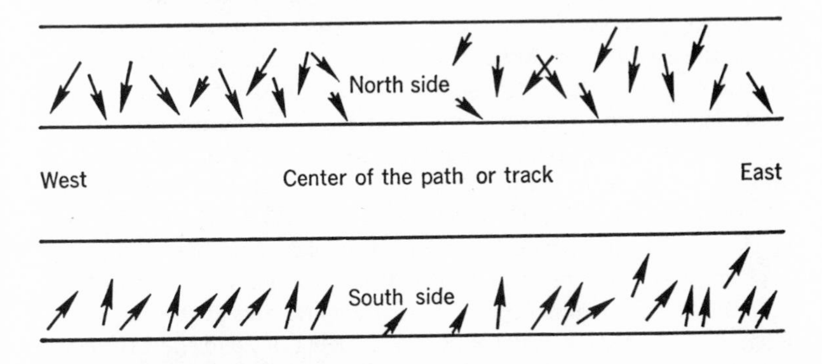

It will be perceived in this case that all the lines of prostration on the southern margin of the track will, if produced, cross the center of the track to the *eastward* of the original position of the tree, and seldom or never to the westward thereof. The same is also true of a large portion of the trees which fall on the northern side of the track, and for an obvious reason which I shall again allude to; but there is still found a considerable number of trees on the northern side, the lines of direction of which, if continued, will cross the center of the path to the *westward* of the original position of the fallen tree.

It is on this important distinction that the whole question which has been agitated at Philadelphia, appears to turn; for if the wind from all sides of the tornado moves or blows directly towards its center, then, as Mr. Espy has finally admitted and now claims, the effects on the two opposite sides of the path must precisely correspond with each other. But if trees on one side, point inward towards the rear of the tornado, and those of the other side also point inward, but towards its *front,* it follows that the wind does not blow towards the center of the tornado but in a whirl or circuit, winding no doubt spirally inward and upward at the same time, in the manner of all ascending vortices.

The reason why no greater portion of the trees on the north side of the track are found inclining to the westward appears to be this. The specific motion of the whirl *on this side* if it turns towards the left is more or less retrograde as regards the line of progress, and the progressive velocity of the whirl is therefore abated from the force of the wind during the earlier part of its action, in passing off, the specific direction of the whirlwind has here become Northwesterly, *combining the relative velocity with both the inward and the progressive motions;* and the trees which had momentarily withstood the more diminished action of the first portion of the tornado, are now prostrated.

The theory which supposes the wind of the tornado to blow in the direction towards its center, sufficiently accords with the position of the fallen trees as I have found them on the northern side of the track, so far as this point alone is concerned. But if the theory be true, *there should be found as many trees inclined westerly on one side of the track as on the other,* which cannot be the case with a whirlwind, for the specific direction of the latter on one side of the track must *coincide* more or less with its progressive course, excepting only as it is modified by its inwardly spiral and at the same time ascending motion, the latter movements being of course always commensurate with the activity and violence of the whirlwind.

I trust that these reasonings, which are perhaps out of place, will have no influence in diverting your mind from the facts which I wish to obtain, for I have often congratulated myself with the idea of finding in you a correct observer for that portion of our country where the occurrence of these tornados is not infrequent. That your official duties necessarily make large demands upon your time I need not be told, but the bearing which the facts in question have upon correct views in meteorological science will, I have no doubt, be deemed a sufficient apology for instituting the inquiry,

I am with much Respect
your obedient servant
Wm. C. Redfield

P.S. Your reply to this inquiry will be expected with some interest, and I shall be glad to receive permission to make use of the same if I should find occasion.

yours WCR

Loomis Papers, Yale University

Loomis to Redfield

Western Reserve College October 22, 1838

Dear Sir

Your acceptable letter of Oct. 9th arrived a day or two since having been first sent to Hudson, N.Y. and there remailed. I was particularly pleased to learn that there was a possibility of obtaining additional information respecting my favorite storm through Colonel Reid,[35] and I have accordingly written to him informing him precisely what I want. As I do not know his address I enclose the letter to you requesting that you would direct and forward it. I have just seen announced a book by him entitled "an attempt to develop the Law of Storms," royal 8vo price 1 £. 1 s. published by John Weale 59 High Holborn. Have you seen the book, what do you think of it. And can it be obtained in New York? If it contains anything new, I shall wish to obtain it forthwith. Perhaps you may have seen Col. Reid and can inform me who he is. I learn that he made a communication to the British Association at the last meeting but cannot learn very definitely what it was. Being so far from New York, I find it difficult to obtain regular information from Europe. This is one of the greatest inconveniences of my position and I feel it sensibly. It takes a great while to get a box from New York, as it is very liable to be mis-sent or to be cast aside in some ware-house on the route. I thank you for the additional extract from your own meteorological Journal. Perhaps you think I make slow progress in collecting information and that it is high time for me to publish. I have collected [a] great deal of information, having barometric observations from the two stations and thermometric from nearly a hundred stations within [one word missing] of the storm. But I wish to tell the whole story at one breath and shall therefore wait as long as there is any prospect of obtaining additional information. I think however I shall publish as soon [as] I can hear from Col. Reid.

With regard to the direction of winds at different localities there is a good deal which is obscure to me. These N.E. storms are common I believe throughout the whole Atlantic coast. Here they are almost unknown. Thus I find in the July No. of the Franklin Journal page 37 that *two thirds* of the whole quantity of rain at Philadelphia for three years came from the N.E. I suppose the same is substantially

[35] Sir William Reid (1791–1858), a British army engineer, later Governor of Bermuda. He and Redfield exchanged data and became collaborators although they never met.

true of the whole Atlantic coast North of Philadelphia if not also South of it. Here we occasionally have rain from the N.E. but very seldom, and the wind is sure soon to get rains to the N. or N. West; or sometimes to the S.E. and South. It is the comment of New England people here that they have entirely escaped the N.E. storms of the coast. My impression is that the same is true of the Western States generally, and that those frightful N.E. storms do not extend beyond the Alleghany mountains. Here then is a problem to solve. The N.E. storms of the Atlantic do not depend upon latitude but upon *distance from the coast,* and it will naturally occur to everyone that this direction is the *direction of the coast,* I do not [c]learly comprehend the physical cause of this phenomena. It is [a] point upon which you have of course reflected and I would like to hear the results at which you have arrived.

My thanks are due to you for proposing me a member of the N. Y. Lyceum and to the Society for electing me. Whatever fee there may be for admission I shall cheerfully pay when informed what it is. I do not know how extensive is the *plan* of the Lyceum, yet you are probably aware that I have never devoted my attention to Natural History, technically so called. In Meteorology as in all the branches of Natural Philosophy I feel a deep interest. I perceive from Silliman's Journal that the Lyceum has published three volumes of Annals and a part of the fourth. I have the impression however that they are almost exclusively confined to Natural History technically so called.

What has become of the Albany Institute? I cannot learn that they have published any thing new for a long time. Are they defunct or quiescent? I have understood that Mr Webster their former Secretary had been unfortunate in business and was obliged to abandon entirely scientific pursuits. Is it so? The number of scientific men in this country is so small that we feel the loss of a single one. We are now in a new phase of weather. During summer it was remarkably hot, and the month of September was unusually clear and pleasant. Thus far in October we have had little but rain, clouds, and raw winds. Saturday October 6th it rained from 5 P.M. to 8½ P.M. amount .406 inch. Oct. 9th from 3 to 4 P.M. .026 inch; Oct. 12 8 A.M.—4 P.M. amount .223, Oct. 14th. during preceding night to 3 P.M. .675. Oct. 18th. 8 A.M. to 19th. 10 A.M. .871. Total 2.201 inches in thirteen days. But the provoking part of it is that it does not become *clear* when it ceases raining; but the clouds continue until another [r]ain, and that too whether the barometer be high or low. On Wednesday morning Oct. 17th 9 A.M. the barom. stood at 29.339 inches the highest [I h]ave ever observed at this place, the lowest was Feb 16th. 1 P.M. 28.086., the greatest range yet observed 1.253 inch. This embraces the last ten months only. I am satisfied that there is more cloudy weather in Hudson than in New Haven. When a North West wind here succeeds a warm rain [w]e expect dense black clouds for at least a day and often longer [a]fter the rain ceases. This I apprehend is to be ascribed to the influence of the Lakes. I hope to hear from you fully and frequently on the subject of meteorology as well as other subjects. I prize highly scientific [co]rrespondence as it is now my only means I have of communicating with [sc]ientific men.

With much respect I am yours truly
Elias Loomis

Redfield Papers, Yale University

Redfield to Loomis

New York October 30th 1838

. . . The Lyceum of Natural History considers itself constituted under the general sense of the term or that of Natural Science, although its published annals have consisted chiefly of papers on Natural History, technically so called, to which indeed its proceedings have a general leaning. Should we ever reach the point in this city, when separate organizations can be advantageously made of the votaries of science, the Lyceum would probably then become more strictly limited in its objects. At present it covers, in this city, the whole ground, there being no other scientific association. The Lyceum is of long standing, and its published papers have been creditable to the science of the country. The stated fee of corresponding membership is five dollars, which was imposed many years ago, in order to prevent an improper enlargement or selection of its correspondents. . . .

Loomis Papers, Yale University

Henry to Redfield

Princeton April 15th 1839

My Dear Sir

I have just finished reading the article in the Edinburgh Review on the labours of yourself and Col. Reid and although the *"hour has reached in nights black arch the key stone"* I cannot refrain from stealing a few moments more from sleep to congratulate you on the highly complimentary tone of the review and the clear and interesting manner in which your labours are thus placed before the world.

You have long since felt the consciousness of having arrived at an important truth but this was not sufficient reward for your labours; this truth required to be appreciated and acknowledged, by what is considered, high authority abroad, in order that it might be received with confidence by the public generally and thus be placed on the way of practical application. That you may long be enabled this to add to the honor of our country and to advance the cause of science and humanity is the sincere wish of

Yours
Joseph Henry

Redfield to Henry

New York June 3rd 1839

Professor Joseph Henry
Dear Sir. Your esteemed favor of the 15th ultimate was duly received and I owe you my acknowledgements for the kind and flattering manner in which you speak of my investigations in the matter of the storms.

Although the subject has attracted but little notice from abroad within the last two years, yet the manner in which it is now received and treated by some of the most distinguished votaries of science in Europe, cannot but be grateful to my feelings. The Article in the Edinburgh Review to which you allude was written, according to my private advices, by Sir David Brewster. An article on the same subject from Sir John Herschel [36] was also expected in the London Quarterly Review for April, which owing to some cause does not appear. I received however a few days since, a polite letter from Capt. Basil Hall,[37] accompanied with a copy of an article on the same subject from the Foreign Quarterly. In this article the Captain

[36] Sir David Brewster (1781–1868) was a leading British physicist and the editor of a leading scientific journal. Sir John Herschel (1792–1871), the son of Sir William Herschel, continued many of his father's interests and had many scientific contributions to his credit. Redfield had arrived.

[37] Hall (1788–1844) was a Navy captain who had made himself unpopular in the United States with a frank account of his travels here. He was known for his pendulum observations of the force of gravity, a favorite geophysical endeavor.

does much to show the practical application and importance of our newly acquired knowledge of the law of storms. I have also just seen a portion of another review, in the May No. of the United Service Journal, from an unknown hand. [Capt. Smyth R. N. Astronomer etc.][38]

It is a matter of sincere regret with me that our friend Mr. Espy is not content with his own position, but thinks it necessary to renew his attacks upon my statements and conclusions relating to the storm of September 1821, as you may have seen in the March No. of the Journal of the Franklin Institute. To this I have felt it necessary to reply and to speak with some freedom as well as candor of his general course of proceedings in this matter, all which I was desirous to avoid.

I am indebted to you for a copy of your late paper on electricity, which I have read with much interest and trust that you are pursuing your inquiries in this important branch of science. I am interested to know if any thing yet discovered will show any minute relations of the electric action with the involute rotations of a small whirlwind. In the magnificent electric developments which are often made in storms I cannot yet persuade myself that electric action stands in the relation of *cause* to the attending phenomena, but rather in that of an effect; notwithstanding the opinions expressed by the reviewer in the United Service Journal. But if a spiral coil or helix produces such strong inductive effects in electric experiments, why not the rapid and extensive intertwining of heterogeneous[39] layers of atmosphere in the rapid involutions of a powerful whirlwind produce analogous effects and on a grand scale? This is a matter on which you can best form a correct estimate. I am with much respect, yours very truly

Wm. C. Redfield

Redfield to Espy

New York 17th November 1839

Dear Sir

Your favor of the 5th instant came to hand this morning. It would give me pleasure to examine the track of the Newark tornado at the time you propose, but my engagements for the week are such as may

[38] Redfield's bracketed insertion.

[39] Heterogeneous as regards temperature and hygrometric and electric inductions. [Redfield's explanation]

prevent it. I have no doubt that the appearances at Newark are essentially the same as in the track of the New Brunswick tornado, and others which we have severally examined. I have never supposed, however, that the movements within a vortex or in the interior of tornados, were *"at right angles to the radii of a circle," as you seem to infer;* and this error I suppose to be one of those which lie at the foundation of your inductions.

To establish the bare fact of *"the translation of storms in space and their continuity in time,"* you are doubtless aware, is little more than to complete a work that was commenced by Franklin nearly a century ago. Our personal interests or claims in these matters are, however, of small moment, and must await the awards of the future.

Although my own investigations have led me to results which differ essentially from your views, yet I shall rejoice at any actual advance which you may accomplish in meteorological science. I am dear Sir very Respectfully yours

Wm. C. Redfield

Charles Darwin to Redfield

12 Upper Gower St London Feb. 24. [1840]

Sir

Having been much interested by your paper on "Whirlwinds excited by Fire" which has been reprinted in the Edinburgh New Philosophical Journal, I venture to call your attention to a published account of the effect of a volcanic eruption, which may possibly have escaped your notice. When the Island of Sabrina was formed off the Azores, Capt. Tillard (Philosophical Transactions of the Royal Society 1812 p. 152) describes an immense body of smoke rising from the sea, which when "in a quiescent state had the appearance of a *circular cloud revolving* on the water like an horizontal wheel in various irregular involutions etc. etc." He then adds "the cloud of smoke now ascending to an altitude much above the highest point, to which the ashes were projected, rolled off in large masses of fleecy clouds, gradually expanding themselves before the wind in a direction nearly horizontal, *and drawing up to them a quantity of water spouts, which formed a most beautiful and striking addition to the general appearance of the scene."* This part appears to me singularly interesting in relation to what you have said in Art: 11 of your conclusions. Taking your account of the whirlwind produced by artificial fires, we here see the very cause of one set of waterspouts. This case appears to me the more important, because the inference, deduced by Prof. Oersted[40] from the accounts which he has compiled of waterspouts and whirlwinds (a translation of whose paper appeared in the number (53) of the Edinburgh New Philosophical Journal, previous to that which contained yours) seems to be that the cause lies in the currents of air flowing in opposite directions but following a parallel course. In the case of Sabrina Island or rather Crater, we see several waterspouts resulting from the whirl produced by great ascending currents of steam and gasses.

I trust you will excuse the liberty I have taken in addressing you: having witnessed waterspouts on the Coast of Brazil. I have al[ways] felt a great curiosity to understand their origin, and therefore could not resist the temptation of being possibly *instrumental* in adding a singular fact to the data which you already possess and which you have brought to bear in so admirable manner on meteorological phenomena. With the highest respect

I have the honor to remain
Sir
Your obedient servant
Chas. Darwin

Darwin to Redfield

Atheneum Club [Dec. 1840]

Dear Sir

I have been prevented by long continued illness from not having many months since sent you my sincere thanks for your most valuable pamphlets on meteorology (which, however, I do not feel worthy to receive from the little attention I have paid to that great branch of science), and the Geological Reports some of which I have been able to peruse and have been much interested by them. As I am yet far from recovered I beg you will excuse the lateness and briefness of this letter and believe me dear Sir

Yours very faithfully
Charles Darwin

Post[s]cript.

With respect to the latest discussion on the *rotary* action of whirlwinds, I will just mention a *trifling* observation I noticed in

[40] Hans Christian Oersted (1777–1851), the Danish scientist better known for the discovery of the deflection of a magnet by electric current.

Bruce's[41] travels, when he describes the sublime appearances, presented by the great whirling columns of sand on his return home across the Nubian desert. He especially says that the *whirling movement* had left traces or concentric furrows (?)[42] on the pointed conical hillocks of sand, with which the plains in parts were scattered. These hillocks having been left, where the columnar mass of sand broke. I quote only from memory.

P.S. 2 As I have mislaid your letter, I am compelled to direct this to the care of Prof. Silliman, who, I daresay, will excuse the liberty.

Redfield to Darwin

New York May 1841

Dear Sir

This will be handed to you by my friend John Blunt Esqr. of this city who I beg leave to recommend to your acquaintance.

Mr. Blunt has kindly offered to hand over to you a copy of the last geological report made under the authority of the state of New York. It is less ample than the previous reports of progress which have been made in the course of the survey, and will probably be succeeded in the course of next year by the full and final report.

The late meeting of the Association of American geologists held at Philadelphia on the first week in April was attended with a fine spirit and feelings of deep interest on the part of its members. Some expectations had been indulged by gentlemen engaged in the surveys of our extensive Silurian formations, of meeting with Mr. Murchison[43] on this occasion.

I venture to send you also a few copies of my short paper on the tornado which passed across the State of New Jersey in June 1835: being nearly the same as published in January number of the London and Edinburgh Philosophical Magazine. As this tornado was, by experienced seamen who saw it pronounced to be a veritable *water spout,* I have thought that a set of determinate observations upon its whirling action might not be unacceptable to some of your

[41] James Bruce (1730–94), a noted African traveler.
[42] Darwin's question mark.
[43] Sir Roderick I. Murchison (1792–1871), a noted British geologist.

naval or scientific friends. I have on hand several other cases of this kind which have not been printed.

I am dear sir very respectfully yours
Wm. C. Redfield

Redfield Papers, Yale University

THE WILKES EXPEDITION

❂

Before the Civil War put an end to a golden age of clippers and other less romantic but capable sail-propelled craft, the sea was an American frontier like the prairies and mountains in the West. American merchantmen and whalers carried the flag around the world to little-known seas and lands. There was considerable American interest in the South Seas, still largely unexplored. The fur seal fishery early spurred an interest in the lands below Cape Horn. One of the fur sealers, Nathaniel Palmer, went far enough south in 1820 to earn a claim as a discoverer of Antarctica. There were good practical reasons for America to want exact information about regions promising wealth to its citizens.

Without the stimulus of scientific curiosity, however, and the desire for national prestige from scientific accomplishment, it is doubtful that Charles Wilkes would have sailed out of Hampton Roads on August 18, 1838. The principal propagandist for the United States Exploring Expedition (to give its formal title), which Wilkes headed, was Jeremiah Reynolds. His attention was first called to the Antarctic by the theories of John Cleves Symmes (1742–1814) which postulated a hollow earth with openings at the Poles. After going south in 1829 with a private expedition, Reynolds returned cured of the Symmes theory but infected with enthusiasm for a great national expedition. Both he and his supporters in this proposal were well aware of commercial opportunities. But as an amateur naturalist, Reynolds was primarily attracted by the scientific challenge in exploring little-known lands and of the great prestige America could reap. Not that the idea was original; the American expedition would be following the wakes of earlier British, French, and Russian expeditions. In 1836 Reynolds' skillful blend of science and commerce resulted in legislation authorizing the new nation's greatest scientific venture.

Over two years elapsed before the ships sailed, a delay nearly fatal to the expedition. The government had little experience in outfitting expeditions of the scope demanded; Secretary of the Navy Mahlon Dickerson proved indecisive, if not incompetent and hostile; and the naval officers were often more zealous in matters of protocol than of logistics. The first appointed commander, Captain Thomas ap Catesby Jones, despite the prodding of a letter from President Andrew Jackson of July 9, 1836 (which appears below), could not overcome the only slightly concealed opposition of the Secretary and of a good many Navy officers and resigned in November 1837.

While the naval end of the expedition was conspicuously lagging, Jeremiah Reynolds, now corresponding secretary of the Expedition, was recruiting a very strong scientific corps mainly composed of young, promising scientists. In the letter of November 16, 1836, to the Secretary, which appears below, Reynolds outlined his views on the objects and conduct of the expedition, drawing freely upon the advice given to the Navy Department by various individuals and scientific societies. Three aspects of this letter are noteworthy: Reynolds' strong plea for the support of science; the suggestion of conflicts between the naval officers and the civilian scientists; and his emphasis on natural history over geophysics. The last two were to prove fatal to Reynolds' connection with the expedition. Even before he wrote to the Secretary, Lt. Charles Wilkes, the head of the Depot of Charts and Instruments—the seed from which the Naval Observatory grew—was sent to Europe to purchase instruments for the expedition. To the great disgust of the naturalists, Wilkes brought back only astronomical instruments. As events proved, Wilkes had different ideas about the role of the naval complement.

With Jones out of the picture and with Dickerson clearly unable to organize the expedition, the new President, Martin Van Buren, placed the entire matter in the hands of the Secretary of War, Joel R. Poinsett (1779–1851). A cultivated amateur naturalist, Poinsett had experience in the War Department as an administrator of land-exploring parties. Stressing the purely scientific nature of the enterprise, he passed over officers with greater seniority and selected Wilkes, a very junior Lieutenant, as head of the party because of his known scientific interests. Things moved quickly now, but not wholly as planned by Reynolds.

In the 1838 memorandum of Wilkes to Commodore Chauncey which appears, in part, in this section, geophysics supplanted natural history as the principal object of the expedition. In consequence Wilkes goes on to propose that the civilian corps of scientists (a great source of difficulty to the naval officers) could be replaced by the medical officers as far as possible, apparently since the civilians' duties were less important. Poinsett, mindful of the need for the support of the scientific societies, overruled Wilkes. But the geophysical work was confined to Navy officers (meaning Wilkes), and Prof. Walter Johnson of the Franklin Institute, originally selected for these fields, was summarily left behind.

Wilkes wanted the Navy to have the prestige of science as did the Army Engineers, who were busy surveying the West. Or as he put it in a letter to the Secretary of the Navy of July 16, 1842 (see p. 123), the Navy should not "act simply as the hewers of wood and drawers of water" for the scientists. To the scientists who had already observed the protocol squabbles in the Jones period, Wilkes' willingness to subordinate science to the advancement of his service's prestige was undoubtedly disturbing.

In this period many civilian scientists began to harbor doubts about the military services as patrons of science, doubts which would persist in one form or another until the present.

Wilkes returned from his circumnavigation of the globe anything but a hero. He was immediately faced with a court martial on charges brought by his subordinates. The English explorer Captain Sir James Clark Ross denied that Wilkes had found land in the Antarctic at the reported latitude and longitude. Most serious of all the blows was that no one was prepared to do anything with the mass of data brought back by the party. Eventually volumes of reports came out under the auspices of Congress, but in limited editions of 100 copies so that few scientists could read the results. Most of the volumes appeared so long afterwards that their contents were stale. By and large, the natural history volumes are rated now above the geophysics volumes. Wilkes, though a tactless and contentious person, deserves considerable credit as an efficient and brave commander and for his unremitting efforts over more than twenty years to getting the reports of the expedition through the press. After his return from sea the United States did far better in organizing scientific expeditions, many stimulated by the example of this first great American scientific venture overseas.

President Jackson to Secretary of the Navy Dickerson

Washington July 9th 1836

My dear Sir,

About to leave the City for a short time, and feeling a lively interest in the exploring Expedition directed by Congress and more particularly from the great solicitude expressed by all the members of Congress that it should be sent out as early as possible, and more particularly as the Executive is anxious that nothing should be wanting on our part to secure its success; and if unsuccessful that no blame should rest upon us. It is my desire that ample means as authorized by Congress be furnished, and prompt measures taken to prepare and compleat the outfit. To effect these objects: Let Captain Jones be informed while he will be responsible for the due execution of the project of the enterprize, that the preparation of the means, the selection of the officers and agents etc., and a general superintendance of the outfit under the Secretary of the Navy will be with him. That these views may be carried into full effect, I desire, that the Secretary of the Navy order Captain Jones to repair to the Norfolk and New York Stations there to confer freely with the commanding officers of those stations touching the arrangements and equipments of the Ship and other vessels which are to compose the expedition; and that Captain Jones may be instructed to enter at once into this business and that Captain Jones with such other officers selected for this expedition, as may be necessary, be ordered to open endeavours for recruiting their expedition crews.

That harmony may exist, I would suggest the propriety that no officer should be selected to whom Capt. Jones may have well founded objections. With these arrangements, should the expedition fail, the responsibility will rest with Capt. Jones, and not with the Department.

I am with great respect
your most obedient servant
Andrew Jackson

P.S. The Secretary of the Navy will see that this is wrote in great haste, much blotted, and having no time to copy, and having great anxiety that this expedition should be fitted out in such manner as will redound to your credit as the head of the Navy. A.J. It will be proper that Mr. Reynolds go with the expedition; this the public expect. A.J.

Wilkes Expedition Letters, National Archives

Torrey to Henry

New York Sept. 26th 1836

My dear friend

I have been chasing about the country so much since I received your letter of August last that I have hardly time to reply to it. If possible I will endeavor to write several letters for Bache if he has not sailed. Pray let me know.

Have you heard of the great national voyage of discovery for which preparation is now making? There are several vessels to be sent out by our government in about two months, a frigate and 3 or 4 sloops of war, under the command of Capt. Jones. There will be a scientific Corps, the members of which will be well paid. Pickering and probably Dr. Gray will go. It is supposed the salary will be 3000 dollars a year. My object in writing now is to ascertain whether there is any probability that Alexander could be engaged for the expedition? The situation and salary would be made very desirable, but I fear his mathematical engagement will be a barrier. He will have (if appointed) the best instruments that can be obtained for astronomical, magnetic, and meteoric observations. This is all between *ourselves* at present, but if he says *yes* we can put things in a train to consummate his wishes. The vessels will be absent 3 years. Perhaps, however, you know all about the affair, but I wish an answer as speedily as possible, for if Alexander can't go, they will look out for some other person. They are determined to have a *big team!*

I have been working on the state survey and am quite tired of travelling, but I explore about here and work at home. It will not be in my power to visit Princeton this commencement. I must preside at a pretty important meeting to be held tomorrow evening, and on Wednesday evening also (D.V.) I shall be occupied. I regret that I cannot be with you, but so it is. Pray come here this vacation and let us have a long talk. I wish to know all your plans and feelings and to tell you mine. How are college matters prospering?

My very kind regards to Mrs. Henry and all your amiable family, and in these Mrs. T. joins. Ask Mr. Wiggins if he will be so good as to bring on the apparatus for condensing carbonic acid gas [one word unclear] I left in one of the closets near the door.

I suppose he will remain a little while at Princeton yet. Remember me to him.

Yours truly
John Torrey

Historic Letter File, Gray Herbarium, Harvard

Reynolds to President Jackson

City of New York Nov 16th 1836

Sir,

Since I had the honour to receive from you my appointment of corresponding Secretary to the Surveying and Exploring Expedition now being fitted out under the direction of the Navy Department my attention has been constantly and steadily directed to the substance of the main objects of the enterprize, especially as regards the civil or scientific corps, upon the organization of which the successful issue of the undertaking particularly depends. The selection of the most competent persons for this province being a matter of such great importance as well as peculiar difficulty, I have deemed it my duty to pay particular attention to the subject.

The various departments of scientific investigation clearly seem to arrange themselves under the following heads, viz—

1. Geology and Minerology.
2. Botany and Vegetable chemistry.
3. Zoölogy.

The other sections of scientific and philosophical which will require to be provided for are—

1. Meteorology.
2. Magnetism and Electricity.
3. Astronomy.
4. Philology.

The important fields of Geology and Minerology may, it is thought, be entrusted to one person, although if the limits of the corps will allow the addition, an assistant or companion would be desirable.

The more extensive department of Botany, including whatever relates to its technical or descriptive part, and also to vegetable production in an economical or commercial point of view, is of primary importance. So far as descriptive Botany is concerned, the duties must necessarily devolve chiefly on a single individual, because the subject does not admit of those convenient natural divisions which, in Geology, render the branches far more independent objects of pursuit. From the comprehensive field embraced by the science in question, it seems, however, requisite that some eligible person be associated with the botanist, to devote exclusive care to its practical applications in furthering the ends, and increasing the economy of commerce; and that, for this purpose, he be charged to investigate the probability of naturalizing desirable vegetable products of the countries explored, in the southern portion of our own soil. The plants of one island, or section of continental land, are often found to flourish, when introduced beneath distant climes, to a much greater extent than is commonly imagined. It is a remarkable fact, well known to botanists, that nearly all the plants of Japan with which we are acquainted, find in this country a soil and temperature peculiarly congenial to their growth. The same may be expected as regards the other islands on both sides of the tropics, and, if experiment confirm the supposition, important accessions to our sources of national wealth and com[fort] will be thus placed within our reach. When it is recollected that the British Government once dispatched an expedition expressly to introduce the bread fruit into their West Indies colonies; when we consider the probability that the plant which yields that valuable material, New Zealand flax, with others of equal utility, would doubtless thrive in Florida and our Southern States, the value of these considerations will perhaps be appreciated. They cannot how[ever] be fairly estimated except we bear in mind the vast variety of climate contained within the limits of our latitude and longitude, from Maine to Florida from the Atlantic to the Pacific, naturally adapted to the growth of almost all the useful vegetable productions of the earth. It should also be remem[bered] that the diversity of organized substances which form the basis of our manufactures and commerce, and on which all our resources depend very few are the spontaneous harvest of our own land.

The range of Zoology is so immense, especially in the reg[ions] contemplated to be explored by the expedition, as to require the combined [atten]tion of several persons, among whom the different branches should be distributed, according to the tastes and acquirements of the individuals selected.

That no portion of that wide field may be left unprovided for, it is necessary that the distinctive predilections and attainments of candidates sh[all] be first consulted, so that an harmonious and efficient Zoological corp[s] may be formed from the best available materials. I have no doubt that this delicate and responsible task may be so accomplished as to sec[ure] the talents of men of science, willing to embark in the enterprize, in those branches in which they are most eminent; and, also, so to combine their ef[forts] as to give assurance to the government and the public that the duties of each subdivision of this important department will be satisfactorily and ably performed.

Meteorology, Magnetism, and Electricity, with those contingent considerations relating to climate may be comprized in the duties of a single department; if so, care should be taken that the individual appointed be competent to investigate phenomena in each of the above named sciences.

Astronomy may perhaps be added to the Hydrographical division. Should this be the case, I think it will be better to confide it to the nav[al] officers.

I must now beg leave to submit a schedule, which will be framed both with reference to organization, and the nomination, of persons for the different situations; so as to agree, in substance, with the reports transmitted to the Secretary of the Navy by scientific societies, in compliance with his request for information on these subjects. It will at least be found to embody all those views in which there is a general agreement. Unimportant particulars may differ, and this is of course to be expected in a matter where several bodies act independently of each other.

1st Minerology and Geology,	to be placed under charge of James D. Dana professor in Yale College. Two in this department are desirable.
2 Botany,	Dr. Asa Gray of New York, with one companion to take charge of Economical Botany and vegetable Chemistry
3 Zoology Mammology,	or the department of Zoology referring to quadrupeds
Herpetology,	Reptiles
Icthyology,	Fishes.

to Dr Charles Pickering, Philadelphia

Ornithology	Birds
Lepidoptera	Butterflies
Taxadermy,	or the preparation and care of specimens
Molusca Nuda Zoophites, Omstaceology,	

to Dr Reynell Coates of Philadelphia

Comparative Anatomy,	
Conchology and Entomology,	to be provided for by suitable distinct appointments or divided among the Zoologists already mentioned.

As regards the qualifications of the above named naturalists, the President is respectfully referred to documents and recommendatory letters in possession of the Navy Department. The divisions of Zoology may not [be] such, exactly, as the gentlemen indicated would choose to arrange among themselves, but the whole matter can safely be left to their discretion, animated, as I know they are, by an ardent zeal, and feeling, as they do, the necessity for that harmony and concert of effort, by which alone their own reputations can [be] enhanced, and the great objects of the expedition effectually secured.

The next matter of interest is the department of illustration. I[ts] high importance will be at once seen and acknowledged. Natural His[tory] drawing, particularly, is of the utmost consequence, as well as by far the most difficult province to fill; requiring as it does a kind of talent little cultivated in this country. Still it is probable, persons maybe found capable of executing this description of drawing, under the direction of naturalists, in a creditable manner. They alone are qualified to dec[ide] on the merits of artists of this class, and it is respectfully suggested, th[at] they be authorized, immediately on their appointment, to make the [ne]cessary inquiries, and that they recommend to the Secretary of the Navy such candidates as they approve.

As too much care cannot be bestowed on this subject, it [is] to be desired that artists, selected for landscape and portrait painters sho[uld] be skilled in Natural History sketching; so as to be capable of assisting Natural History draughtsmen, when consistent with their regular employment; or occupying their places should the latter be temporarily incapacitated.

In view of these considerations, I would strongly recommend Mr. J. H. Shegogue, of New York, as portrait painter, (a situation he is calculated to fill with credit to the expedition) not only on account of his proficiency in that branch of his art, but also for his acknowledged talent for copying animated nature generally. In this respect he is highly spoke[n] [of] by the naturalists of New York who have inspected the works of his pencil and I understand the Lyceum of Natural History here, has commended him, on the above grounds, to the favorable notice of the Secretary of the Navy. I have no doubt the nomination is, not only judicious, but the very best that could be made.

Mr. Raphael Hoyle of Hudson N.Y. is mentioned in language of high praise by the most distinguished artists of the country, nor is it claiming too much to say, that he ranks only second to Mr. Cole among the landscape painters of the United States. I shall consider this branch well bestowed if com[mitted] to his hands. The testimonials in possession of the Navy Department wil[lfu]lly confirm what I have said with reference to this gentleman.

Prof. Walter R. Johnson of Philadelphia, of whose attainments it is unnecessary to speak, has I believe signified his willingness to take charge of Meteorology, Magnetism, etc. He would prove a valuable acquisition to the scientific corps, being unquestionably qualified to conduct the researches in this department to the satisfaction of the government and the public.

Regarding the choice of philologist and ethnographer, I have a few words to say. No European expedition as far as my knowledge stands, has been accompanied by an individual in this capacity. The benefit to be derived from such an accession is forceably pointed out in the letters of the Hon. John Pickering and Proffrs. [Charles] Anthon and Gibbs,[1] contained in the pamphlet I had the honor to forward you. From the candidates who may apply, I hope no one will be selected without the approving testimony of Messrs. Dupanceau of Philadelphia, Pickering of Boston, Anthon of New York, and Gibbs of New Haven, who are among the few scholars in the union competent to give a just opinion on this point.

I have been notified that Prof. C. Pentice, of Geneva College has declared his readiness to join the enterprise; my inquiries respecting his attainments have been met with the universal response, that he was a right and accurate scholar, peculiarly fitted by his studies for Philological investigations. I deem it highly desirable that his services be accepted.

The interests of the expedition emperiously demand that the chief appointments in science should be made as early as possible, not only that the naturalists may resign any situations that they may now occupy in our various institutions, but also that they may prepare and procure such apparatus etc. [as] can only be provided by those who are to use them in their separate departments. With the selection of the gentlemen enumerated, the corps might be considered embodied, the announcement of which fact, joined to the official recognition of the naval officers, will be received with delight throughout the nation; the time of sailing is a matter of much less moment. It should be immediately on the completion of the preparations, not an hour before.

There has been much discussion on the subject of compensation among the members of our literary and scientific associations, and I hazard nothing in stating the unanimous opinion that each gentleman belonging to the scientific board should [tear, one word missing] three thousand five hundred dollars per annum, with stores found. This is [the] first occasion ever afforded to the Executive of this country to fit a value on science, by [co]nceding to its profes-

[1] Josiah Willard Gibbs of Yale is better known in the history of science as the father of the mathematical physicist of the same name.

sors a fair equivalent for their exertions. The effect will be for good. In this age of sordid speculation, it will be a noble record, one that w[ill] tell well in history, that mind as well as matter has received its just appreci[ation].

The gentlemen whose names have been given, are precisely the individuals on whom the Executive may safely rely for all reasonably to be anticipated in [the] wide range of inquiry before them: their appointments as designated, would n[ot] only produce concert of action, but are I know the very arrangements they a[re] solicitous should take place.

It has been intimated, that the naval officers would object to these civilians (if so they may be styled) receiving so large a sum as three thousand five hundred dollars per ann[um]. I have heard no such expression of feeling from those with whom I have conversed, nor do I see on what ground such objections can be sustained. The labours of the two classes are as distinct as their professions, a[nd] can have no more bearing upon each other, as regards their respective remunerations in this Expedition, than has the pay of a naval commander on the salary of a professor in one of our universities. Besides it should be borne in m[ind] that the gentlemen of the Navy will continue to receive their incomes after their return while the members of the scientific corps will have to commence the world an[ew].

In the plan of organization here given, I do not wish to assume any cre[dit] to myself. I have been merely a close observer of what was in progress, a[nd] most ready and happy to accord to the Secretary of the Navy all due praise for pains he has taken to procure from learned societies, as well as from men of scientific eminence individually, the aid of their counsel in all duties apertaining to the civil portion of the enterprise.

In this communication, I have felt it a duty to limit myself to what has been recommended by those societies and gentlemen, at this instance both with regard to the number, and names of the persons I have mentioned. Amidst the numerous and conflicting claims which, I am aware, have in some cases been pressed on the attention of the Secretary, I trust he will truly estimate the motives which prompt me to take the liberty of addressing you; and do me the justice to believe, that no sentiment of disrespect to him are among my [tear in text]. . . .

Wilkes Expedition Letters, National Archives

Extract from Wilkes' Memorandum, 1838 [2]

11[3] Scientific Dept—All the duties appertaining to Astronomy, Surveying, Hydrography, Geography, Geodesy, Magnetism, Meteorology, and Physics generally to be exclusively confined to the Navy officers, these are deemed the great objects of this expedition and it is co[nfi]dently believed that there are some who are so well qualified to perform them.

The other scientific Depts. consisting of Zoology, Geology and Minerology, Botany and Conchology, it is proposed to fill up as far as can be from among the medical corps that will be attached to the Expedition; if however none of the medical officers of the Navy can be found sufficiently qualified or willing to undertake to become principals in any of these Departments, then I would suggest that there be appointed the following, viz—

Two persons for the Zoological Depts.
One " " Botany
One " " Geology and Minerology
One " " Conchology

And two draughtsmen well qualified in all the Different depts of drawing, which number of persons are the most that can be accommodated on board the vessels without great inconvenience to their officers and crews. The assistants in all the above departments to be exclusively taken from among the Medical and other officers that may be attached to the Expedition and their tastes and qualifications may point out. All the appointments that may be made to fill the above scientific situations, to be Naval appointments so as to place them entirely under the control and direction of the Commander of the Expedition.

12 The foregoing arrangement and organization will beyond a doubt be the most efficient for the Expedition giving ample employment and usefullness to all, making the service desirable and sought after by the officers, useful to the Navy, honorable to the

[2] Sent to Commodore Isaac Chauncey, president of the Board of Navy Commissioners, a body that administered many professional naval activities in the period. Chauncey forwarded the memorandum to President Van Buren who transmitted it to Secretary of War Poinsett.

[3] The first ten paragraphs of this memorandum dealt with nonscientific matters, that is, outfitting and administration of the expedition.

Country, and highly advantageous to the commercial interests in its results as well as to science . . .

16 I deem it highly essential and necessary to the character of the Expedition that a series of Experiments should be made with the Pendulums, and magnetic Instruments prior to the sailing of the Expedition. Not only that the Instruments may be again tested and results had but that the officers who are to act as assistants should become somewhat acquainted with the duties apertaining to them. The Results will serve as comparisons to future Experiments made on the return of the Expedition, which will show what dependence is to be placed in the results had during the voyages; Washington is deemed by me the best place for these Experiments on various accounts (which it is needless here to mention). Some of the Instruments I believe are now here, and the last can be brought from New York; among the things that will still be required is a small portable house for the making of the Pendulum Experiments and also for the protection of the Instruments when in use from the weather, as many officers as will can be spared from the duties of the vessels will be employed as assistants in making the Experiments. If the weather should prove favorable the Experiments can be completed within a fortnight.

17 An arrangement will be required to be made with the Messrs. Bond and Sons of Boston for making observations of the Moon culminating Stars and other Phenomenon during the absence of the Expedition. Similar rearrangements will be made for these observations at the Depot Naval Instruments on Capital Hill by the officers attached to it, and this is the site I should select as the most fit for making the observations and Experiments prior to sailing. . . .

Extract from Poinsett's Endorsement of Wilkes' Memorandum

The first part of the 11th Mem. is concurred in but it would appear injudicious to dismiss entirely the whole of the Scientific Corps. The number designated by the Memorandum may be found too small to be carried into effect without creating much clamour, but every judicious effort ought to be made to reduce the minimum as low as possible. The propriety and even necessity of doing so the different literary and philosophical Societies may not only acquiesce in but will probably lend their aid in accomplishing . . .

Their assistance may be taken from the Medical and other officers of the Navy and the Civil Corps may have either such temporary appointments or be so subjected to the rules and regulations of the service as may place them under the direction and control of the Commander of the Expedition. . . .

Joel R Poinsett

Wilkes to the Secretary of the Navy

U.S. Flag Ship Vincennes
Harbor of Callao July 1st 1839

Sir

I have the honor to submit the following report of the operations of the Exploring Expedition under my command from the 25th of February last to the 30th June, having forwarded in my dispatch No. 25 the report of our operations from Rio de Janeiro to Cape Horn including the Rio Negro.

Captain Hudson sailed from Orange Harbor near Cape Horn, with the Peacock and tender Flying Fish in company, on the 25th February in pursuance of my instructions, copies of which were forwarded to you with my dispatch No 22: and proceeded to make an exploration of that part of the Southern Ocean comprised within the 105th degree of West longitude and the Western Coast of Palmers land.

A few days after sailing the vessels were separated in a severe gale, and proceeded to the rendezvous designated in their orders in the event of a separation, but they did not again meet until the 25th of March when after having narrowly escaped being frozen up and imprisoned by the ice, they were proceeding northward, having encountered severe weather and ascertained the utter impossibility of penetrating further south . . .

We [the *Porpoise* and the *Sea Gull*] sailed from Orange Harbor on the 25th of February with a fair wind, and on the 1st of March encountered a few ice islands, reached King Georges Island, one of the South Shetlands which we found completely covered with snow; on the same day we made Bridgeman's Island and prepared to land, but were prevented by the heavy fog which enveloped us in the course of fifteen minutes: its whole appearance was volcanic with smoke issuing from its cliffs and the atmosphere impregnated with sulpher to the leeward.

From Bridgeman's Island we steered a direct course for Palmers land for the purpose of tracing its eastern coast as far as possible, this being, (as I had previously informed you) the only feasable operation in my opinion at this late period of the season.

We made Palmers land on the morning of the 2nd of March toward the S-West and encountered very many ice bergs of all dimensions and much drift ice discovering as I then thought the projecting head land; stood on in hopes of finding an open passage along and near the rim [?], but on approaching it within three miles of the Cape or head land, found a continued pack of ice islands which completely blocked up all access; some of the islands appeared to have been but recently detached from shore: the land was seen standing to the Southward and Eastward for a distance of 15 or 20 miles.

Off the cape and near our position we discovered three small rocky islands with icebergs attached, one of which we passed within a few hundred yards, but the ice prevented our landing.

As the wind was leaving us and the ice islands drifted by the current and wind upon the land and towards us, it was found necessary for our safety to extricate ourselves as soon as possible, we therefore commenced working the Vessels out towards the North and eastward through much drift ice, and continued nine hours endeavoring to obtain a position we deemed sufficiently safe during the night, which was very dark with much rain and fog: we were at this time surrounded by upwards of 80 ice islands within a circuit of five miles.

My next endeavour was to pass around the main body of the ice and effect an entrance farther to the Southward, in hopes of getting a sight of land farther Southward and eastward in the direction it was seen tending. We bore away next morning at day light (March 3rd) and steered to the Southward under a press of sail among icebergs some of which were from two to three miles in extent. The weather continued wet and stormy with much fog which had been the case, excepting for a few hours from the time of our leaving Orange Harbor.

While off Cape Hope Palmers land when we obtained observations for chronometers and latitude by double Alidade which placed it in latitude 63° 10′S. and longitude 56° West.

Considering it to be too hazardous to remain during the night among so many dangers, I hove to and at day light on the morning of the 4th I continued on course Southerly but could discover no opening to enable me to close in with the land to the Westward.

During the night the weather continued with snow and fog, temperature of air being 32°, water at surface 30° and at 100 fathoms depth 29°.

Constant observations with Sisc's self-registering thermometer of the temperature of the Sea upon the surface and at the depth of 80 to 100 fathoms were made as the best guide to prevent our being frozen in: On entering among the ice islands the mercury had stood at 34° at the depth of 80 fathoms and fell gradually afterwards to 28½ degrees.

The wind which had been fair, changed from North to S.S. West and increased to a severe gale with snow, sleet and hail; the temperature of the water at the depth of 100 fathoms had fallen to 28½ degrees and the air to 27° on the morning of the 5th.

It became now difficult to work the Vessels, owing to the quantities of ice which had collected on them with no prospect of a change in the weather or wind, the salt water at a great depth had fallen to the temperature of freezing and much ice around us, I concluded after consultation with the Commander of the Porpoise to retrace our steps to the northward before the wind, and increase our distance from the land to prevent being frozen in . . .

Wilkes Expedition Letters, National Archives

Wilkes to the Secretary of the Navy

Washington City July 16. 1842

My dear Sir

Agreeably to your desire I hasten to give the information relative to the remaining duties of the Expedition and that are *absolutely* necessary to carry out the instructions of Congress in passing the act authorizing the Expedition Viz. for the promotion of the great interests of Commerce and Navigation and to extend the bounds of science and promote the acquisition of knowledge.

For the accomplishment of these great objects there was required persons to attend to the different departments of Science and the following was the organization which I proposed and was adopted by the government, and the most economical one that could have been arranged to carry out the great views intended, and that the accommodations of the Vessels would permit. Viz. The departments of Astronomy, Hydrography, Magnetism, Meteorology, and Physics including the experiments with the Invariable Pendulum was confined to myself with the officers under my command as assistants besides the above I was charged with the History or Narrative of the voyage.

This at once greatly reduced the Scientific Corps, which had been organized Viz from 23 to 9.

I felt that the Navy was justly entitled to all these departments embraced as they are within the limits or scope of the profession, and that they ought not to be, attached to such an undertaking to act simply as the "hewers of wood and drawers of water" as was the case in its original organization.

{ Charles Pickering { Titian P. Peale	Naturalist
Horatio Hale	Philologist
James D. Dana	Geologist
William Rich	Botanist
[W.D.] Brackenridge	Horticulturist and Asst. Botanist
{ Joseph Dayton { Alfred Agate	Artists
J. P. Anthony	Conchologist [4]

formed the nine to these was added a mechanic for the repair of instruments and their proper preservation.

In all the above departments much remains to be done. Indeed I view the service of the above as necessary now and even more so than at any other period of the cruize nor can their services be dispensed with or the work concentrated without great loss to the Expedition and the reputation of the country.

For my own departments I require the services of Mr. Stewart who was a clerk in the Expedition but whom I have made Hydrographical Draughtsman and some few of the officers, who have been my principal assistants. Mr. Stewart will be enabled to assist me in my copying etc. etc. He is one of my own scholars and is now engaged in the duties assigned him.

I truly regret that anything should have occurred to dampen the ardor of those who are attached to the Expedition and absolutely necessary to the bringing out the results the ardor that has been felt during the cruize has been all important to our success

[4] Mr. Anthony was with the Expedition until the end of Nov. 1839 after which period his duties were divided among the rest and successfully performed. [Wilkes' note]

and has been in every way encouraged by me. And I did hope that it would have been kept alive until all had been accomplished. The reputation of our Country is at stake and if what has been attempted and succeeded in, is not now finished from any notion of economy, or derangement of the organization all will be ruined and we shall become the laughing stock of Europe and all the praise that has been lavished on our Government for its noble undertaking prove but satire in disguise.

What will be the reputation of those who have had the ordering of things since its return on this becoming known on the other side of the waters? For the reception of myself I can easily account, but that of the officers and crews is truly unaccountable, particularly the want of any expression of thanks to the latter on their discharge. It was felt by every officer and remarked by every man. On the minor duty I have been gratified by it formerly, and I have with pleasure seen its effects upon many of the men that formed a part of the crews of this Expedition when on other service with me. I have urged it all in my power but without effect, every day develops some new opposition to the Expedition. I am aware you think I want cause for this opinion perhaps I am mistaken but I cannot but feel myself bound up in it. Indeed it would be strange if I was not and I must say it is heart sickening to me to hear those who have shared in its dangers and troubles complaining of a want of attention and courtesy, and exhibiting the unceremonious discharge from their duties, with little or no prospect of consumating the labors, in which they have been engaged for the last 4 years. And before they have even seen their friends; some are suffering under sickness contracted from their exposure in the service of their country. They are now suddenly cut off and destitute of support for themselves and families; these facts are well known. Such treatment is without a parallel in the service of this or any other country.

Contrast our Expedition with those of the French and English engaged in the same service and at the same time, honor and rewards are heaped on all before their return. Examine our results, compare them with theirs, contrast us in every way with them you please, or with Expeditions that have gone before us, and then ask if we have not reason to feel mortified. Do not misunderstand me, I do not ask anything for myself, and will not as long as this mist hanging over me exists, which any fair and candid examination into my actions and conduct would have long since dissipated. Neither do I ask impossibilities or undeserved praise no greater punishment can be inflicted on the head of one who receives it, but I would ask is it not fully apparent and placed beyond cavil that the men of the Expedition have done their duty and did deserve the thanks of the Department before they were disbanded. It was openly complained of when they were paid off.

C. W.

Wilkes Papers, Library of Congress

THE GEOPHYSICAL TRADITION II: HENRY, BACHE, AND THEIR CIRCLE

Dispatching a great exploring expedition was not the only sign of a rise in scientific activity in the United States as the decade of the 1830's drew to a close. A whole new generation, as it were, of scientists had appeared. These new men were pursuing their preferred lines of research, pressing for more science in the colleges and universities, and, in general, seeking to advance scientific research in the United States. Practically every major scientific field of the day was involved in the flurry of activity.

One group of scientists achieved a position of dominance in the American scientific community in the two decades preceding the Civil War. Intellectual attainments, force of character, and occupancy of strategic institutional posts—all combined to give them considerable influence. The membership of the group, known as the "Lazzaroni" (or scientific beggars), was not fixed nor was it formally constituted. By mail and at festive meals at meetings of the American Association for the Advancement of Science, jobs were parceled out, new organizations were launched, and national policies were decided upon.

Joseph Henry and Alexander Dallas Bache were the leading spirits. Eventually Benjamin Peirce, the Harvard mathematician, took a prominent part in the deliberations of the circle. Most of the members were deeply involved in geophysics. A naturalist like Louis Agassiz or an experimental chemist like Wolcott Gibbs were exceptions. From time to time, an influential layman might get an invitation to a dinner. But in the main the Lazzaroni were professional physical scientists, mostly interested in geophysical problems, who admitted a few kindred souls from other fields to their ranks. Their interests and range of influence extended to all of the sciences and included much of the research performed in universities and the government. They were consciously promoting the development of a professional scientific community in America.

What happened in the physical sciences in the United States from 1840–60 is largely a story of the growth of institutions, programs, and research ventures of this circle. To a much lesser extent, this is also true of the natural sciences. With a few crucial exceptions, physical scientists outside the group and its allies were minor figures of little influence on the course of development of the sciences in America in this period.

One stranger to the group who clashed significantly with them was Matthew Fontaine Maury, the oceanographer. Documents of Maury's or about him appear in this section because the conflict with Maury was important and also illustrates some of the obstacles encountered by the Lazzaroni. Perhaps the most significant physical scientist omitted from the circle in terms of later developments was the chemist John William Draper[1] of New York. He had few contacts and no dramatic struggles with the Bache-Henry circle. While they exercised dominion over much of the scientific community, Draper and similar investigators in Europe were quietly establishing lines of research which would eventually eclipse the scientific interests of the geophysical tradition. Although generally underrepresented in the ranks of the Lazzaroni, natural scientists had contacts and alliances with the circle, especially in the conduct of explorations and surveys. Conspicuously absent was a good friend of Joseph Henry's, the botanist Asa Gray.

The Search for a National Weather System

A continuing concern of American scientists of the geophysical persuasion in the first six decades of the past century was the establishment of a national meteorological system.

The field seemed at a crucial state of development; with judicious support, important theoretical generalizations and valuable practical applications appeared almost within grasp. Several members of the Lazzaroni were important participants in the movement to make the weather a subject of scientific study under the auspices of government.

Scattered observations of the weather at given points on the earth's surface were not enough; to understand and to predict simultaneous observations were needed. As early as 1814 the Army surgeons at military posts were directed to keep meteorological diaries, a practice that continued through most of the century. For a brief period starting in 1817 the General Land Office required its field offices to keep weather records. In 1825 New York State formed a weather system in which Joseph Henry participated. A Pennsylvania system came into existence in 1839 with James P. Espy as the principal meteorologist.

Although Redfield sharply differed with him and Loomis was probably more sophisticated scientifically, Espy in the 1840's occupied the center of the stage as the "Storm King" and the great promoter of national weather activities. His writings, especially the book *The Philosophy of Storms* (1841), gave him an international reputation. Shifting base to Washington in 1842, Espy became in effect the national meteorologist.

[1] For Draper, see pp. 197–98, 252–58.

Until his death in 1860 his friends in Congress kept him on the public payroll by tacking his funds onto appropriation bills for the Navy, the War Department, the Patent Office, and any other available place. From the network of observers he organized, Espy received and published compilations of meteorological data.

When Joseph Henry came to the Smithsonian Institution in 1846, he planned and eventually inaugurated (1849) a system of telegraphic reports of weather conditions. Espy's network of observers were gradually absorbed into the growing Smithsonian system; Espy himself, while cooperative, never fully gave up his role as a free agent of meteorology. Henry had the support of most of the leading meteorologists as well as the public—by 1860 about five hundred stations reported meteorological data to the Smithsonian. As early as 1849 or 1850 Henry started displaying daily weather maps. Our present Weather Bureau grew out of the Smithsonian's system.

Espy to Loomis

Washington City April 30th '42

My Dear Sir,

I fear you will have thought me very negligent in not answering your interesting communication of the 4th March for so long a time. That communication was never received till yesterday. I hasten to say in answer that I am much pleased that you are so zealously engaged in the investigation of Storms. Much good to science, and of course to mankind will be the result of the combined labors that are now brought to bear on this interesting subject.

I send you herewith all the documents which I have been able to collect on the storms which you are now investigating: you will please to return them to me as soon as you can conveniently. Enclose to the

Navy Department
Washington City
Care of J. P. Espy

and thus my letters will come free. I hope before long to have the aid of Government, in the further prosecution of the most important branch of science. I am sure of encouragement from you. If you have not published your investigation of the late tornado, perhaps you could send the Manuscript on immediately to me, I shall be in Washington for three weeks yet and then I go to New York and Boston for a short time, but I shall probably be located in Washington before long. The information which I have received concerning the chickens and turkies remaining alive after being stripped of their feathers in a tornado is verbal and I have no doubt of the fact though I cannot refer you to the authority by name. A gentleman told me that he saw turkies walking about naked after the passage of a tornado which occurred many years ago; but he added that they soon died. This was the first intimation I ever had of the facts and he told it to me as a strange phenomenon, which came under his own observation. I know not how to explain it. There is no doubt also of *persons* having been stripped entirely naked in the same way without being *hurt seriously*—I have the evidence in my possession, and will send it to you if you wish it.

Please to give me a list of all the persons in the U.S. and Canada etc who, you think, will keep journals of the weather for me, and I will send them blank forms when I begin operations again. I wish to have a large number of simultaneous observations made over a wide extent of territory, give me your aid and advice and encourage your representatives in Congress to do what is right in the premises. I thank you for the work you have already done to advance Meteorology; your name will be handed down to posterity as one of the pioneers in this beautiful science, and one Storm which will always bear your name, will be quoted as a perfect specimen of inductive philosophy.

I hope to hear from you soon, in the mean time I remain with respect yours very truly

James Espy

Until the Adjournment of Congress you may endorse your letters to the Hon. H. W. Beeron M. C.

If I receive any more letters on the storms in question I will send them to you. What formula do you use for correcting Bar. for Capillarity and temp?

I have not answers for any of your questions not noticed.

Espy to Loomis

Washington City May 22nd '42

Dear Sir.

I have received yours of the 17th Instant in which you acknowledge the receipt of *one* letter from me containing documents etc. Now I wrote to you three different times, sending documents every time. I hope you receive these documents *free,* as they are franked by a member of Congress.

We have had here a N.E. and N.N.E. wind all day the 20th and

21st without rain except on the morning of the 20th till about noon. This morning the wind was S.S.W. gentle with a few drops of rain about noon. Now at 5 P.M. the sun is beginning to shine for the first time for those 3 days.

The wind was quite strong on the Morning of the 20th. Barometer rising all day up to 30.37. It is now 29.94 and falling fast.

I shall probably not want my documents returned until you have finished your investigation of the storms, which have undertaken, and it gives me great pleasure to be able to facilitate your object, for your object and mine is the same, to discover *truth* on this highly interesting department of science. There is enough for all of us to do. The whole scientific world will feel grateful to you or what you have done already on this subject, and I most sincerely hope that you will have entire success in your attempt to investigate the storms of Feb. What do you think of my attempts to answer the various objections which have been used against my theory or Philosophy of Storms? If you have any Objections to my theory in whole or in part please to state them either to me or to the public. I wish all my doctrines to undergo the most rigid examination by the brightest minds our country can boast.

My errors will then be detected and what is true will stand out in bolder relief.

I should like extremely to know all the phenomena attending this last storm of the 20th and 21st Instant.

I think it quite probable that it travelled towards the south, and that when it quit raining here with a strong North or N.N.E. wind continuing, it was still raining in N. Carolina and S. Carolina, and that the wind at Pittsburg and probably as far as your place was N.W. If I get 300 observers, what a power we shall have to answer such questions as these!!

Will a ventilator not costing more than ten dollars sell to a great extent in your country, if it cures all chimneys which smoke from the wind, and free all privies from bad smell? I have invented such a one.

If you answer this letter immediately I will receive your answer before I leave Washington.

Encourage all your correspondents to send journals to me directed War Department, Meteorology, and I will receive my letters *free*

I remain very truly and affectionately
your Brother in Meteorology
J. P. Espy

Espy to Loomis

Washington Jan 11th '43

Dear Sir,

I am glad you have found time to examine my calculations in my "Philosophy of Storms." You have detected one mistake. It originated from not attending to refraction and the length of time the semidiameter of the sun takes to sink below the horizon. I thank you for this correction. I hope most earnestly that you will find time also to examine the fundamental points in my theory, which will be found in the first Section, and also more briefly from page 304 to page 311. I ask your particular attention to these last pages.

Prof. [Benjamin] Peirce in his "Review" of my work commenced in the Cambridge Miscellany has made a most important mistake, which when corrected will turn the whole calculation in my favor. He takes it for granted that if air goes up to where it is under one tenth less pressure it will expand into one tenth greater space but this it will only do provided the temperature remains the same; but as the temperature falls in its upward motion and contracts in bulk on this account—it will not expand in proportion to the diminution of pressure. (see page 32 of my book) Again the degree of cold produced by a given expansion depends in some measure on the temperature, the higher the temperature the greater the cold.

Again the degree of cold by a given expansion is affected by the density of the air, the greater the density the greater the cold. ceteris paribus. All these causes combined will cause a much less degree of cold as the column of air ascends to greater heights than Prof. Peirce seems to admit. Let me know what you think on these matters; at your earliest convenience. On last Saturday night commencing at 11 o'clock and all Sunday morning we had a strong wind about South or perhaps S. b. E. with but little rain. In the evening it cleared away with gentle breeze S.W. and bright sun-set. Bar. lowest 29.83 two or three hours in P.M. about 4^h & 5^h. I think it not improbable that you had a westerly wind at the same time, and that there was a storm of some severity on the lakes.

Do not take patterns by my short and unsatisfactory letters. You have no conception how constantly my time is occupied. If I neglect to answer any of your questions it is because I have no satisfactory answers, nevertheless if you reiterate any question I will say something about it, to show you that I notice your requests.

I wish also you would find time to examine the evidence which Reid's storms afford of an inward motion of the air, and let the world know your opinion.

What is your opinion as to the course of storms? Do examine again if necessary and let me know if the air could get in on one side more easily than on the other of the late tornado. Have you no difficulty in getting the air supplied from a narrower space on one side of the centre of the path than on the other? What do you think of the article that appeared in the Globe of the 7th Dec. on Meteorology? I hardly expect you to answer all these questions unless I were more particular in answering yours. I hope if there is a meeting of Scientific men we shall see each other and then I will expect a long conversation etc.

Yours very truly J. P. Espy

Espy to Loomis

Washington City Feb 16th '43

Dear Sir,

I have had a conversation with the Secretary of the Navy on the Subject of the Observatory, which you enquired about in your letter to me. He thinks nothing will be done for some considerable time, different from what is now doing, and he says it is probable that officers in the Navy alone will be employed in the Observatory: but nothing final can be determined at present.

You are probably aware that they are now making magnetic and meteorological observations in a temporary observatory. I shall always attend to any suggestions you may have to make on the subject.

Can you lend me your journal of the weather for 1840? I suppose it contains barometric observations.

It may be enclosed to the

Navy Department
Washington City

with the word "Meteorology" on the envelope, and can be returned the same way.

I remain very truly yours
J. P. Espy

Is it not remarkable that Mr. Redfield in his controversy with Prof. Hare persists in the assertion that storms travel towards the N.E. when you show that your great storm commenced at Quebec about the same time it did at C. Hatteras? and sneers at me on the same subject? Speak out on this to the world. Moreover Mr. Redfield acknowledges that a fall of the barometer accompanies the progress of storms. Now you have shown that the barometer fell 3 hours sooner at Montreal than at Flushing!!! and some 50 hours sooner than at Bermuda. How can men permit themselves to be so blinded by preconceived opinions?

Loomis Papers, Yale University

The Rise of Astronomy

Building a great observatory in America was one of the goals of the friends of science almost from the very launching of the new republic. National pride and the progress of science both required a permanent structure equipped with the latest marvels of the lens grinder's skill. Many attempts were made to enlist the aid of both the federal and the state governments, as well as to raise funds by private subscription. Both Philadelphia (under the leadership of the American Philosophical Society) and Boston (through the agency of Harvard) vied for the glory of erecting the first observatory. The most determined try for the support of the national government occurred during John Quincy Adams' administration (1825–29) but foundered on partisan hostility.

All these unsuccessful attempts failed fundamentally for the same reason. The sums required for a proper astronomical institution were simply too large for the America of the years 1790–1830. When observatories at last appeared in the 1830's at Williams College, Yale, and Western Reserve (the last under Elias Loomis), they were modest affairs not comparable to the leading European institutions. But when Loomis reviewed the progress of astronomy in America in 1856, he was able to list 25 observatories maintained by the national government, educational institutions, and private individuals. By 1882 the United States possessed 144 observatories. Some of these were world famous. This growth of astronomic activity was extraordinary in itself but especially for a nation which had rejected support of astronomy in 1825 and was reputedly indifferent to basic research.

The observatories, especially in the first eight decades of the century, served a dual role; they were the physical embodiment of the great achievements of Newtonian science and the laboratories of geophysics. Part of their growth stemmed from little more than the desire of Americans, now more affluent than in earlier days, to copy the appearances of European civilization. Astronomers could get funds from prestige-conscious laymen for observatories; physicists could not obtain equivalent support since their laboratories did not have the same appeal. But many of the observatories were also heavily committed to work in geophysics. Whereas chemistry would rise in America from the stimulus of agriculture, medicine, and industry, physics would be nurtured by astronomy.

The origins of the observatories were very diverse and more often than not the result of chance rather than of rational planning. Harvard, for example, had long wanted an observatory and added William C. Bond (1789–1859) to its staff in preparation for the day when Cambridge, Massachusetts, would have a great telescope. Bond was a watchmaker who had erected a private observatory at Dorchester, Massachusetts. But only when public interest was aroused by the Comet of 1843 could prosperous Boston raise the needed funds. With the public subscription brought forth by the comet, Harvard opened its observatory in 1847. William C. Bond was succeeded by his son, George P. Bond (1825–65). Being very good observers and outstanding for their development of instruments and techniques, the Bonds soon placed the Harvard Observatory in the first rank with their discoveries.

The Naval Observatory had a more involved gestation period. Congress, after all, had on many occasions after John Quincy Adams' proposals expressed firm opposition to any permanent federal astronomical establishment. A few devoted Navy officers, however, neatly steered the observatory past Congressional shoals. In 1830 the Department of the Navy organized a Depot of Charts and Instruments. To test some of the latter —ships' chronometers, for instance—astronomical observations were necessary. After Lt. Charles Wilkes took command of the Depot in 1833, he erected a small observatory building on Capitol Hill at his own expense. When Wilkes sailed on his expedition, he arranged for a series of observations at Washington corresponding to his overseas longitude sightings. These were executed by his successor, Lt. James M. Gilliss, who went ahead and prepared also the first star catalog made in America. More important, Gilliss succeeded in getting an appropriation for a permanent observatory. This was not the first federal observatory; the Army Engineers, the scientific elite of the War Department, had constructed an observatory at West Point in 1840. In two letters to Loomis of July 4, 1843, and October 18, 1844, printed below, Gilliss tells about the origins of the Naval Observatory and of his feelings when Maury was named its head.

With two first-rate observatories in operation, American contributions to astronomy increased. European and American investigators exchanged data, engaged in co-operative ventures, and independently tackled the same problems. Sometimes controversies would arise from these European-American interactions in research. Two such incidents are treated in letters appearing in this section. This first involved two Americans in a dispute, the second placed an American scientist in the awkward position of mildly denigrating one of the great discoveries of Western astronomy in the nineteenth century.

In 1847 Dr. C. L. Gerling of Marburg University, Germany, called for a new determination of the solar parallax, an angle important in the determination of the mean distance of the earth to the sun, which is the unit of measurement of the solar system. Gerling proposed observations of Venus particularly and also of Mars. Gilliss, now working with Bache in the Coast Survey, apparently leaped to the opportunity and exercised his considerable promotional talents in organizing an astronomical expedition to Chile. Some of his maneuvers are detailed in the letters that follow. Particularly noteworthy is Gilliss' account of Maury's opposition in a letter to Loomis of April 10, 1848 which appears below.

Gilliss was in Chile from 1849 to 1852. Before leaving Washington, Gilliss had the Secretary of the Navy instruct Maury to execute observations parallel to those in Chile. Without these no true judgment of Gerling's method was possible. Gilliss returned with data from 217 series of observations. During the three years the Naval Observatory had done only eleven series of observations of Mars (six useless and three partly so) and eight series on Venus (two of which were bad). Gilliss brought back a star catalog and other scientific data, but the expedition failed in its primary purpose of determining the solar parallax.

In 1846 the planet Neptune was discovered. Independent of one another, the Frenchman, Urbain Jean Joseph Leverrier (1811–77), and the Englishman, John Couch Adams (1819–92), calculated the orbit of the then unknown planet from the perturbations observed in the orbit of Uranus. When Leverrier predicted the location of the new planet, the Berlin Observatory did see the planet in the designated area of the heavens. It was a great triumph for the theory of the universe.

In America Sears C. Walker (1805–53) and Benjamin Peirce immediately started a series of calculations on the new planet. Walker found that Neptune was seen in 1795 but classified as a fixed star. Finding other past sightings, Walker was in a better position to calculate the true orbit than Leverrier and Adams, and he came up with results markedly different from those predicted by the two co-discoverers of the planet. Peirce went on to point out that the 1846 discovery was due to the fact that the positions on the predicted and correct orbits were coincident during a period of approximately fifty years. The discovery was an accident proclaimed Peirce; if Leverrier had requested an observation before 1800 or after 1850, no planet would have appeared in the predicted position.

While Peirce and Walker were essentially correct, Peirce characteristically was carried away and overstated his case. Leverrier naturally resented disparagement of his great discovery as is evident from the exchange between him and Peirce printed in this section. Peirce was cast in a role similar to Bowditch's in respect to Laplace. While greatly admiring the Frenchman's achievement, he pointed out errors of detail. Leverrier, a master of planetary astronomy, retaliated by disclosing discrepancies in Peirce's work. Walker, then at the Naval Observatory, was eased out by Maury, a Leverrier partisan.

Gilliss to Loomis

Georgetown D.C. July 4th 1843

Dear Sir,

Yours of the 26th ultimate reached me yesterday. I have intended acknowledging your meteorological Journal for a long time past, but that my whole time has been occupied since last October with matters connected with the new Observatory. I was much pleased to hear your name mentioned in warm terms of commendation whilst I was in Europe, and promised Lamont[2] (at Munich) to send him the Journal you forwarded me. I know that he appreciates it, and will give you the Journal of the German Meteor. Society in return. Pray send him a copy.

With respect to the new Observatory. The Government has only made an appropriation for a "Depot for the Charts and Instruments of the Navy:" but, from a report which I prepared for the Naval Committee in Congress and which they adopted as their own; it was clearly understood, that Astronomical, Magnetic, and Meteorological observations were to be a part of the duties required of the Officers attached to it: were, in fact, essential to the Navy. After long trouble, and vexation, I succeeded in getting the bill passed. It gives us $25,000 for the building and directs the President to give us any ground suitable for the purpose, which belongs to the U.S., and is within the District of Columbia. We have 19 acres. I then pointed out to the Secretary of the Navy how the instruments might be obtained without a special appropriation, the discussion of which might not aid the whole plan: and he directed me to obtain them. In the plan which I submitted to him; I proposed to have a Director who should be a Captain in the Navy, and 4 Lieutenants; (one to each principal inst:) with 8 Passed Midshipmen to attend to the Magnetic and Meteorological Observations and details. There would be no difficulty, however in having any number of Assistants which the Director may desire and my only anxiety is, that an energetic officer may be placed at its head.

Very truly Yours
J. M. Gilliss

I presume there can never be a difficulty about publication: any friend to the institution in Congress calling for a copy of the

[2] Johann von Lamont (1805–1879), a Scottish-born astronomer who entered the priesthood and lived in Germany.

observations, they are furnished by the Department and printed as Public Documents. At all events, I hope to have my labours of the last four years published in this manner. What is to be the fate of the establishment, it is difficult to predict: under an able Director, I am satisfied it would present more and better work to the world, than any Observatory existing.

Gilliss to Loomis

Washington City 18th October 1844

Dear Sir,

Yours of the 14th reached me yesterday, and as I have always credited the adage "short accounts make long friends," I make leisure to give you an early reply. You will have seen, I "guessed" your change of locality from the paper addressed to your new domicile some days since; and indeed I should have strained a point to see Prof. [John William] Draper and yourself, when in New York in Sept, had not all my joints been stretched to the utmost already.

The papers gave correct information respecting the Observatory here. My instructions from the Department terminated with its completion, and although there is no duty so agreeable to me as astronomical labours, yet, I have too much pride to *solicit* a connection with an establishment whose existence is owing solely to myself. You know, the law recognizes it only as a Depot for the Charts etc of the Navy; and Lieut. [Matthew Fontaine] Maury has moved into it, with his Instruments, charts etc. What the intention is, I know not, for I have also too much pride to enquire, although, I believe my head would be brought with sorrow to the grave, if I thought I had laboured to found a mere Depôt.

If it is to be an Observatory: Maury is not the man to be at its head, unless he has an entirely different taste from that induced by his previous life and labours. He is too, physically incapacitated from using instruments, and it is not very probable that the publications in the [National] Intelligencer in March 1843, will dispose the astronomical world to rank him even with Mr. Smith [?].

As soon as I complete my report on the construction, and erecting the building and instruments (which will be by 1st Dec.) I shall take up the observations made in connection with the Exploring Expedition and on which Prof. Bradford has been engaged as my Assistant during a year. I hope we shall complete work during the next summer.

I am very sorry you did not come this way and see our instruments, and Observing Chair for the Equatorial, the last a specimen of American ingenuity. It is so contrived that the observer can give himself three motions while seated in it. However you will see the drawings next winter. . . .

Your friend
J. M. Gilliss

Loomis Papers, Yale University

J. M. Gilliss to Peirce

Washington 22nd October, 1847

Dear Sir,

I take the liberty of forwarding to you copy of a correspondence with Dr. Garling[3] on the subject of a new determination of the Solar parallax. His first letter you have already read, or at least, I sent to you, in the Nat. Intell. It occurred to me that we might have a determination wholly *American;* and I would be not a little proud to encounter the winter of 1849 in such a cause, even on the coast of Patagonia. You will perceive what my proposal is, from letter to Dr. Garling, and the time for preparatory action at home, having now arrived, I earnestly ask your aid in obtaining an opinion or vote of the Academy, to use with the Secretary of the Navy. I know not the proper mode of making such application; and should the annexed petition be improper, or misworded, pray cancel it and present the subject as you deem best. I am sure that you will feel an interest in this matter and that my appeal to your assistance will not be in vain.

Early action is greatly to be desired, as it may be necessary to import a circle, and I am desirous to be ready for Congress when they come, should the Secretary decline the proposal. As soon as I learn your action and that of the Philosophical Society, to which I have written through Prof. Bache, I shall make the matter known to Lieut. Maury and endeavor to prevent his opposition if I cannot enlist his aid. Till then it is to be desired that it should not be discussed. Pray let me know soon.

I wrote [Sears C.] Walker about it sometime since, and have a most eloquent letter from Prof. Bache to whom he had sent my communication. Let W. see the papers, if he is still at Cambridge and say I shall write him in a day or two.

[3] i.e., Gerling.

With my best thanks for the bulletins of the American Academy containing your papers respecting Neptune, believe me.

Your friend & Servant

J. M. Gilliss

Peirce Papers, Harvard Archives

U. J. J. Leverrier to Matthew Fontaine Maury

Paris February 9, 1848

Monsieur et Confrère

. . . I was resolved to be silent on the subject of the strange assertions in America with regard to the theory of Neptune. I cannot, however, permit the two numbers of the *Sidereal Messenger,* with which I was this morning furnished by my honorable friend Mr. Walsh, to pass without a few words in reply. I send them to you, begging that you will give them publicity.

As this is the day for the departure of the packet boat from Paris I have not the time to reply in full; however, what I am about to say will be sufficient to make my justification complete.

Mr. Pierce [sic] has discussed the observations on the satellite of Neptune made by Mr. Lassell, and has found at least nineteen days and a half for the time of revolution. This is a *fundamental error* of that geometer.

The observations of Mr. Lassell assign less than six days for the time of revolution of this satellite. In this they agree with the results of Mr. Struve, which give definitely *five days and twenty-one hours* for the time of revolution.

Everyone will now understand how Mr. Pierce taking such an error as the basis of his calculation on the mass of Neptune, has arrived only at erroneous conclusions. This mass by Mr. Struve's rigorous calculations is equal to 1/14494 of the mass of the sun. I had given it in my work as 1/9322. The *difference is included within the limits* which should result from my theory. It is indeed avowed (page 29, No. 4, *Sidereal Messenger*) that an error has been made when it is said in an editorial note: "Since the above was in type new observations by Lassell and at Cambridge, N. E., on the satellite of Neptune have modified the results; and as matters now stand Neptune may account for 46″ out of 186″ of perturbation known to exist in the place of Uranus." But at the same time it is reaffirmed that the inequalities of Uranus cannot be accounted for by Neptune: "Neptune may account for 46″ out of 186″ of perturbation."

Here is a new theoretical error, The error of 186″ of the old theory of Uranus proceeds from two sources; 1st from the error in the old elliptical elements of Uranus: 2d, from the perturbations, *strictly speaking,* produced by Neptune. When Mr. Pierce says that the perturbations, *strictly speaking,* do not amount to 186″, he says what is true; but he does not tell us *any thing that is new.*

All this is to be found in the work which I have published on the subject. But when Mr. Pierce infers from thence that Neptune ought [orig. cannot][4] to account for the *discrepencies* of the old theory Uranus, he is wrong; because he does not take into account the most considerable cause of these discrepencies [orig. similar][5] that is to say, the inaccuracy of the old elements themselves. Hence it is proved that, up to the publication of No. 4 of the Sidereal Messenger, Mr. Pierce had reasoned from a wrong hypothesis; that he had never recollected that the inaccuracy of the old elements of Uranus ought to be considered.

But in No. 5 there is a change. It is there discovered that the inaccuracy of the old elements of Uranus ought to have been considered; and on page 37 this is presented as an important fact which had been omitted by Mr. Leverrier and Mr. Adams. . . .

They say, page 37, "If he (Leverrier) had lived in that year (1800) and at that time his friend had directed his telescope to the position of the heavens assigned to the hypothetical planet by Leverrier, no planet would have been discovered, and Neptune would have been some 30° or 40° from the computed place of the planet of the theory."

This is false—erroneous. The discrepency here given is four or five times too great, as is the time of revolution which has been assigned to the satellite etc. . . .

National Intelligencer, March 10, 1848

Peirce to Editor, *National Intelligencer*

Cambridge March 13, 1848

Dear Sir

Having just received the number of the National Intelligencer, which contains *M. Leverrier's* letter upon the planet Neptune, I hasten to reply to his strictures upon myself and trust that you will do me the justice to insert my reply in an early number of your

[4] Maury's insertion.
[5] *Ibid.*

journal. He has not touched the main point at issue which is contained in my original communication to the American Academy, including it probably among "the strange assertions made in America," with regard to which he has chosen to remain silent. In whatever light it was viewed by him, it has commanded sufficient respect in Europe to be republished at full length in the most eminent astronomical journal in the world and which rarely republishes anything, viz. in *Schumacher's Astronomische Nachrichten.* It was a correct deduction from M. Leverrier's conclusions, and, bound to his investigations by the adamantic chains of sound logic, could not be overthrown without the prostration of the theory to which it is attached. I confidently reiterate my former assertion that *Neptune is not the planet of Leverrier's theory,* and that it cannot be so regarded even if it should be ascertained that it will account for the perturbations of Uranus. It cannot be included within this theory without so extravagant an extension of his limits, as to destroy all confidence in the elegant analysis, with which he so skillfully contracted them, and which has justly been the wonder and admiration of geometers and the chief glory of his memoir. Having carefully examined the details of Leverrier's calculation, and having followed him through each step of his investigations with ever increasing admiration, I am prepared to maintain that such an extension of his limits is utterly impossible, and that the impossibilty is inscribed upon his formulae in characters which cannot be misunderstood; I am prepared to defend this most ingeneous and most profound theory, even against its illustrious author provided that forgetful, for a moment, of his true glory, he is ready to abandon the substance for the shadow, and to seek popular applause at the expense of sound reputation.

But I have not wandered from M. Leverrier's letter to the main question for the purpose of forcing upon him an unwelcome controversy, and I will, consequently, conclude with a brief consideration of the secondary and comparatively personal points which he has discussed.

And, first of all, I earnestly protest against the proposition that my error in regard to the satellite's time of revolution was *fundamental,* or that it was made the basis of my calculations. The *calculations* were prior to the discovery of the satellites time of revolution. Not more unjustly therefore was the lamb of fable devoured for its alleged disturbance of the waters of the brook which were above it, than am I in danger of being demolished upon the charge of having carried back Lassell's discovery against the stream of time, and made it the basis of calculations which were completed several weeks earlier. I freely, however, acquit M. Leverrier of all disingenuousness upon this point. I am persuaded that he was not aware of the true state of the case, and that he did not know that the error upon which he commented was merely a temporary one, of a single day's duration: scarcely was it made ere it was admitted and corrected: the correction was published the next day in the *Boston Courier,* and I supposed that I had put it into the very same channels of public communication through which the original mistake had flowed. I need not again apologize for it or repeat that it was not unjustifiable, for, great as it was, it was not to be detected by means of the observations which were then in my possession. Its history may teach others the same useful lesson of caution which I have myself derived from it, and which the wisest and most prudent will sometimes require. It was peculiarly appropriate, and perhaps peculiarly deserved, for it resulted from one of those unusual coincidences of which science has other examples, and of which I conceive the discovery of the planet Neptune to be the most extraordinary.

The preceding paragraph disposes of the only portion aspersions of M. Leverrier's letter properly applicable to me, for I do not perceive that he does me the injustice of holding me responsible for the articles in the Sidereal Messenger[6] which are the principle objects of his criticism; although a foreigner may easily forget the vast extent of the American States and might pardonably overlook the fact that Cincinnati is as far from Cambridge as is Paris from Rome. I am not the author of those articles, nor in any way knowing to their composition; I have not even read them, nor any parts of them except the extracts in M. Leverrier's letter. It would be presumptuous in me, therefore, to interfere between them and an attack upon them; and M. Leverrier will on the one hand perceive that his charges of ignorance are wholly inapplicable to me, while the writer in the *Sidereal Messenger* will, on the other hand, not regard me as endorsing the charges by the act of separating myself from the writings, and thus from all necessity of a reply from me . . .

Benjamin Peirce

Text based on draft in Benjamin Peirce Papers, Harvard, and *National Intelligencer*

[6] These were by O. M. Mitchell, the director of the Cincinnati Observatory, who embarrassed Peirce by his injudicious attacks on Leverrier.

Gilliss to Loomis

Washington April 10th 1848

Dear Sir,

I should have answered your last, long ago, and also have acknowledged the copy moon-culminations (printed) kindly sent me last month but, I am sure you will make due allowance for me, when you know the amount of extra writing absolutely necessary in matters relating to the expedition to the southern hemisphere. The obstacles to overcome in obtaining even a *consideration* of it by the Department, have been much greater than I anticipated, owing, it is presumed, to the counsels of Lieut. Maury; for I have been informed by Mr. Stanton[7] (the member having it in charge) that he informed the Naval Committee the expedition was useless, and it is reasonable to suppose he would give like opinion to the Secretary of the Navy. I am happy to inform you, however, that the Naval Committee have *unanimously* directed Mr. Stanton to ask an amendment to the Appropriation bill, and there is reasonable prospect of its success. The report of Mr. S, which (inter nos) was written by Prof. Coffin,[8] will probably be presented to the House today: if not, then on the first day the Committees are called on for reports. It will of course be printed, together with all the letters, and will be sent you as soon as ready for publication. But as a month or two may elapse before the printing is done, owing to prior materials in the printers hands, and the Naval Appropriation bill will come up for action within ten days, in order that we may be supported, it is considered desirable, to have *editorial* commendations of the proposition from the committee, in as many distinct cities as possible. New York being the great commercial city, of course weighs most, as navigation *practically* benefits more from Astronomical observations than any other pursuit; I hope, therefore, you will write for the strongest *democratic* paper in your city (as soon as you receive the National Intelligencer containing the report,) the most laudatory article of the science, patriotic efforts etc. of the party, in first establishing an observatory and now recommending this object, which you can have inserted as editorial. For Mr. Stanton is democrat, and independently of the interest he has always shown in such matters, Bache, Walker and my *Whig* friends deemed it essential to have the proposition from that side of the House. But, I must not tell you more of the secrets

[7] Frederick Perris Stanton (1814–94), a Democrat from Tennessee.

[8] Probably John H. C. Coffin (1815–90), then at the Naval Observatory.

of management here, lest I lessen the awful dignity with which Congress may be surrounded in your mind. You perceive what is to be done, to qualm the consciences of strict constructionists, and encourage the doubtful to do well, and I feel sure, you have too much at heart the promotion of science in our country and the honor to accrue to it for any enterprise having only such objects, not to give cheerfully, all aid. . . .

The full discussion which my proposition has undergone since my last, with Gauss, Encke, Boguslawski, Lamont, Garling, Bache, Pierce [sic], Walker and the two Societies,[9] have resulted in great modification of that previously given you. 1st. The instruments, are to be a meridian circle of the best construction and an equatorial. 2nd. The locality, will probably be in the latitude of Conception on account of climate. (rains.) 3rd. The time is extended to two conjunctions, which with that of going, returning etc. will be near 3½ years. 4th. Magnetical meteorological observations are to be made. . . .

Yours most truly
J. M. Gilliss

Loomis Papers, Yale University

An Enemy of the Circle—Matthew Fontaine Maury

To many of his contemporaries Matthew Fontaine Maury (1806–73) was an impressive scientific figure. He is still much honored as the "Father of Oceanography," although the legitimacy of this parentage is doubtful. But to a large and important segment of the American scientific community, Maury was an outsider, a rival, and a fake. Henry, Bache, and their associates doubted his integrity and his status as a scientist. The roots of Maury's estrangement from his fellow scientists appear in the documents in the preceding section.

Maury was not an astronomer. Yet he had displaced a competent

[9] Maury might describe the expedition as useless, but Gilliss had the weight of scientific opinion behind him, witness this list. Best known of these is K. F. Gauss (1777–1855), the German mathematician. Gauss was director of the Göttingen Observatory and interested in a wide range of astronomical and geophysical subjects. J. F. Encke (1791–1865), a German astronomer, is best known for the discovery of the comet bearing his name, which has the shortest known period of any comet. P. H. L. von Boguslawski (1799–1851) was the director of the Breslau Observatory.

astronomer from the Naval Observatory. His own intellectual interests were elsewhere. Much to the disgust of those who had long agitated for a national observatory, the science languished in his hands. Given the high status of astronomy, this was a cardinal scientific sin. The Harvard Observatory easily eclipsed its federal rival. Unlike Gilliss and other scientifically inclined officers, Maury openly placed the interests of his service ahead of the advancement of science. This, too, was a cardinal scientific sin.

Maury also represented a tradition that was already old-fashioned—that of the amateur scientist—and the Baches and Henrys were the harbingers of the new scientific professionalism. To them Maury's charts were a short-term invention, admittedly a brilliant one, but not a contribution to science. They were appalled by his theorizing in the absence of facts. The patient gathering of data by hydrographic surveys was needed, not the use of unverified figures. His results were useful; they did not, in the eyes of opponents, constitute a science of the seas. Perhaps that only came into existence with the work of the British *Challenger* expedition of 1872–76.

The frictions between Maury and the Henry-Bache circle were due not only to these differences in method and conduct but also to the very great similarities of their interests. Maury was an ardent devotee of geophysics in the service of geographical science. Meteorology is a good illustration of this. Much of Maury's work is describable as the meteorology of the oceans. When he attempted to extend his system to the land, using farmers instead of sea captains, Maury was blocked by the existence of Henry's Smithsonian network. Like Bache and Henry, Maury was an adroit promoter and lobbyist; official Washington and scientific America in the pre-Civil War era were too small to support that many tycoons of science. Although the circle sniffed at Maury's lack of professionalism, the amateur and the polymath were still rampant even in the bosom of the Henry-Bache circle. And lastly, like his principal opponents, Maury had considerable ability and insight, and it was a great loss to science that he and his scientific contemporaries in America could not co-operate.

In one other significant respect Maury differed from the Henry-Bache circle. He was not simply a Southerner but a rabid extremist exponent of that region's cause. Joseph Henry and Alexander Dallas Bache were good friends of Jefferson Davis and other politicians from the South. But they were not committed in politics, only in science. Their scientific work and its results were neutral in terms of any political ideology. Maury was not so inhibited; he dreamed of Southern expansion as far south as the Amazon and promoted scientific work for that purpose. He kept on finding justifications that he called scientific for southern routes for railroads, telegraph cables, and ocean vessels. When the South seceded, Maury defected to the Confederacy. And at long last, Gilliss became the head of the Naval Observatory he had planned and constructed, inaugurating its golden age.

Maury to W. G. Simms[10]

Natl Observatory May 12, 1849

Dear Sir,

I have to day yours of the 5th instant. I owe you an explanation for not addressing you by name in my last for I had mislaid your letter and was not sure that I recollected your address correctly.

I sent the article on the Gulf of Mexico to Bowen[11] and asked him to return it last week provided he would not agree to pay for it, and to publish it in his July No. I impressed this upon him by telling him I attached more importance to its timely appearance than to pay, and that if he would not publish in July I would prefer to let some other journal have it, gratis. I have not heard from him, and I conclude therefore he has decided to accept the article on my terms. He spoke doubtingly about room for it in his July No., and in case he would not make room for it I intended to ask of you a place for the article in your July No. but it appears that you are full also.

I am rather sorry that Bowen agreed to pay for it, because I wrote it with your Journal and Southern readers in my mind. And I offered it to B. to preserve the peace between conscience and inclination, for I felt that I ought to try for pay and having tried and failed conscience could have been quiet. I heard the Southern Quarterly had been in tight places before, and did not much expect that it would be in a condition to pay.

I have seldom or never had the good fortune of having a subject to write upon with which I have been more entranced than with this upon the Gulf of Mexico.

The article on Tehuantepec in your last No., stirred me the more. The writer of that article has evidently not had an opportunity of looking behind the curtain at Moro's survey. A greater imposition was never attempted upon the public, the "Moro" makes 20 feet on the bar of the coatzacoalcos our officers find 12½. He carries 30 feet from the Bar up to Mina-titlan our officers make but 12. He sounded the river when there was a flood in it, and conceals the fact. My main object was to show the commercial resources of the Gulf and Caribbean Sea, the importance of their importance of opening a commercial thoroughfare between them and the Pacific together with

[10] William Gilmore Simms (1806–70), a Southern editor and novelist.

[11] Francis Bowen (1811–90), a philosopher, then editor of the *North American Review,* the leading serious journal in America.

the advantages and consequences likely to arise to this country from such a communication.

In treating the subject it became necessary to consider the extent, fertility, and productions of the river Basins which are drained into this Gulf and Sea, and in doing this the very important fact is brought out that these river basins embrace a greater diversity of character, more degrees of Latitude and a greater extent of country than do all the river basins of the Mediterranean, of Atlantic Europe and India put together. That there the outlet of these river basins are more than 20,000 miles apart requiring more than a year for an exchange of products, that they are all in the same Hemisphere, and have but one harvest during the year. That with us on the contrary our basins are in both Hemispheres with a [reversal] of seasons and consequently with two harvests during the year, and that the commercial outlets to these basins are but 2000 miles apart. And consequently that one ship can fetch and carry as many cargos in a year between these river basins which embrace 70° of Latitude as 10 ships can there between river basins that embrace but little more than one half as many degrees of Latitude.

Only one of the points and you will readily perceive therefor that the subject is both charming and rich.

Uncle Sam works me very hard, and I do not know when I shall have leisure to take up another subject, for I make it a rule never to write unless I have something to write about.

With many thanks for your kind offer,

I remain very truly yours
(signed) M. F. Maury

Maury to Baron von Humboldt

National Observatory, Washington, D.C. Sept 5th 1849

Lt. Maury presents his compliments to Baron Alexander de Humboldt and begs that the Baron will do the Lt. the favor to accept a set of "Wind and Current Charts" which the Lt. has the honor of sending thro' his friend, Professor Rümker of Hamburg.

It is not necessary to allude to the developments of the many interesting physical facts exhibited by these Charts. The Baron's quick eye will perceive them at a glance.

The Lt. may be excused here for alluding to *indications* of certain facts that are being afforded by other Charts of the series which are in process of construction and which of course are not yet published. One of these is a Thermal chart of the North Atlantic Ocean showing from numerous observations made in different years and by a vast multitude of navigators, the observed temperature of the water at the surface for each month of the year.

These records, so far, agree in pointing out a *streak* of cold water, several degrees in breadth, which extends from the vicinity of the Capes of Va. down towards Cape St. Rogue in Brazil. Does this streak of Cold water indicate the existence of a ridge a submerged Mountain range in this direction, on the crest of which, the sea is shoal enough for the waves to bring by their agitation, the cold below, to the surface? The island of Tristan de Achuna, the Penide de San Pedro, Bermuda, seem to stand up as peaks from this sunken chain which formerly connected the two semi Continents, and the mountain ranges of Cuba, [and] others of the West Indies, shoot out like Spurs from the same.

This is all conjecture. A vessel will probably be soon placed under the Lt.'s directions for the purpose of sounding across this streak.

Another Chart is devoted to investigating the migratory habits and places of resort, periodical or otherwise, of the Whale. Right or Sperm.

For this, Abstracts from the logs of a great many whalers have been obtained; the Ocean has been divided off into sections of 5° square, and the Statistics grouped together in such a manner as to show how many days of each month have been spent in each section in search of whales. How many days whales and the kind have been seen, and how many days on which, no whales have been seen.

By this Chart it has been discovered that the Right whale of the South Pacific is a different animal from the Right whale of the North Pacific. The Equatorial regions are to these animals as a sea of *fire*, which neither has ever been known to cross or even to approach within many hundred miles. The Whalemen explain this by stating that the Right whale of the South Pacific, is a small animal, rarely yielding more than 60 bbls Oil: while that of the North Pacific is of a much darker color and is also larger, frequently yielding to the single fish 200 bbls Oil.

The fishermen have their irons marked and numbered, and Right whales, that have been struck in the South Pacific, have been taken in the South Atlantic with the irons in them.

What is the difference between the Whale of Davis' strait and the whale of Behring straits? If they are the same animal would not this fact tend to show that there is at least occasionally a water Communication from Strait to Strait. In other words a N. W. passage?

Respectfully etc. etc.

Maury to Arnold Guyot[12]

National Observatory, Washington
April 8th 1850

My dear Sir

I cannot resist the temptation to express to you the delight and profit with which I have just been reading your admirable "Earth and Man" a copy of which you were kind enough to send me last summer. It is elegant.

The investigations in which I have been so laboriously engaged touching the winds and current of the sea, bear upon many points which you have touched in that capital lecture. You could not in many of your illustrations have been more true to nature if you had had, which you could not have had, the results of these investigations before you.

The U. S. Schooner Janey [under the command of] Lt. Walsh, you recollect was sent to sea last fall for the purpose of making observations in connection with my researches on the water, in the water, under the water and in the air.

She was pursuing a most interesting series of observations, when alas she found herself in such a crippled condition on her arrival at the Cape de Verds as practically to destroy her usefulness to me.

Though Lieut Walsh had not reached the most interesting field for his observations, he has obtained results of high interest, e.g.

Thursday Novr. 15th at 2^h 55^m P.M. Lat. 31° 59′ N. Long. 58° 43′ 25″ W. he sounded up and down and without getting bottom, to the depth of 5700 fathoms or 34,200 feet.

Being provided with the means, he was directed to observe the specific gravity of sea water at the surface and various depths. This he did reducing the water always to the temperature of 60°. Here are some of his curious results, distilled water being taken as the unit

[12] Arnold Guyot (1807–84), a Swiss geographer brought to America by Louis Agassiz. His and similar books were symptomatic of the geophysical trend in America. He went to Princeton in 1854. Guyot was a close friend and collaborator of Joseph Henry.

Nov.[13]	Lat. N.	Long. W.	temperatures surface	depth 100 fathoms
4	30° 46′	67° 34′	1028.3	1028.0

It remains to be seen when Lt. Walsh returns whether he has and I hope he has, preserved specimens of the waters which gave the specific gravities.

But you will observe that in the western half of the Atlantic e.g. to the West of 40° he always found the water from below, when brought up, relieved from pressure and reduced to surface temperature, to be uniformly lighter than the surface water, and on the other side of that Meridian he found it under like circumstances uniformly heavier.

I am opposed to hasty announcements and therefore forbear until Lt. Walsh returns, to commit myself upon the subject. But suppose his specific gravities to be reliable, how important is not the discovery, the clue to which is thus placed in our hands? The data then for a perfect theory as to the general system of aqueous circulation in the ocean will be complete.

Again thanking you for your admirable course of Lectures believe me my dear sir to be very

truly yours

(signed) M. F. Maury

Maury to Louis Agassiz

National Observatory Washington
July 29, 1850

Dear Sir

[I hope to receive before you receive] this some specimens of whale's bone's from the Pacific. They are described thus:

A slab of bone taken from a large r[igh]t whale in the Arctic Sea Lat 69° 40′ N. Longi 168° West,—the whale from which it was taken yielded 180 lbs of oil this however is not the largest I have seen.

Also three small slabs taken out near the throat where it is found much shorter than at the middle of the scalp or jaw. These slabs being attached show how they are bedded in the gum and the distance they are placed apart. Taken by Ship Formosa off Island of Codiac N. W. Coast. Also crown from head of a South Atlantic rt whale (1838) and presented by H. Taber the owner.

[13] Remainder of table omitted.

You will notice the barnacles some of them very perfect.

This crown is placed on the top of the head about 1 foot [from] aft the nose end.

I have the promise of a bit of the crown of the polar whale by Captain Wyatt. I understand that the polar whales have no barnacles.

Shall I send you these bones for Examination and if Yea when.

With thanks for the Kindly consideration in sending me your valuable contribution to the Christian Examiner

I remain yours truly
(signed) M. F. Maury

Naval Observatory Records, National Archives

The Coast Survey, the Smithsonian Institution, and the Lazzaroni

Bache came to Washington in 1843 as the successor to Ferdinand Rudolph Hassler (1770–1843), the Swiss-born scientist who was the first head of the Coast Survey. His good friend Joseph Henry wrote a fine recommendation for him. The Coast Survey then was a small organization not in the good graces of either Congress or the Executive Branch. From Hassler, Bache inherited a set of high scientific standards and the small business of the weights and measures. Before the Civil War, Bache had won the support of officialdom to the level of $500,000 a year—a figure not reached by the geologists until several decades afterwards. The work of the Coast Survey was justly highly regarded in the United States and in Europe as a fine example of the union of pure and applied science. Bache's organization was heavily committed to geophysics—excursions into natural history were very rare. It was undoubtedly the largest employer of mathematicians, astronomers, and physicists in antebellum America. From the vantage point of the Survey, Alexander Dallas Bache could promote and stimulate scientific work in the United States.

Bache wrote one of the recommendations that helped bring Joseph Henry to Washington as Secretary of the Smithsonian Institution in 1846. The Englishman James Smithson had left $500,000 to the United States to found an organization "for the increase and diffusion of knowledge among men." From 1836, when the last Smithson heir died, until 1846 the government could not make up its mind on how to carry out the donor's intent. Among the proposals offered were the establishment of a national university, the erection of an observatory, the endowment of popular lectures, the founding of a natural history museum, and the development of a great library. The 1846 legislation creating the Smithsonian Institution was a compromise and vague on many points. But the library adherents had scored important gains, and the concern for the collections brought back by the explorations and surveys resulted in the inclusion of a museum function.

Joseph Henry's conception of the Institution was quite different from this and was probably incomprehensible to many of his contemporaries. He wanted to use the Smithson bequest to conduct and support original research and for the publication of the results of such investigations. All the other proposals he regarded with suspicion as raids upon the bequest, and he devoted the rest of his life to defending the Institution against its misguided friends. At first the library was the greatest threat, but by 1854 Henry had removed Charles C. Jewett, the Assistant Secretary and librarian of the Smithsonian, the leading advocate of the library concept. In 1866 he transferred the Smithsonian Library to the Library of Congress, laying the foundation for the future greatness of that organization.

The museum was another matter entirely. For many years Henry successfully avoided museum entanglements. But two factors were against him. First, there was the problem of the proper care of the national specimens brought back from the West and from overseas. Second, in 1850 Spencer F. Baird (1823–88), a fine naturalist, became Assistant Secretary and soon was the unofficial co-ordinator of the natural history activities of the government. (Baird succeeded Henry as Secretary in 1878.) Natural history was eminently respectable and the most extensive scientific activity in America. In 1857, Henry consented to the transfer of the older collections for various expeditions to a "National Museum" under his care. From that date on the history of the Smithsonian Institution can be described as a frittering away of Henry's conception and the conversion of the Institution to a museum. The remaining activities in research became largely overshadowed by the care of the contents of what was to become aptly described as the nation's attic.

Henry did support much research, as did his successors. Even more important, the Smithsonian in its early years stimulated scientific work by its publications of research results. For decades, the Smithsonian was a symbol of science in America. It was the one place to which young men interested in science could write and get advice, suggested readings, sometimes jobs, and even equipment. But even during Henry's lifetime, it had acquired the status of a venerable symbol. With this aura of antiquity, strange in an institution so relatively young, the Secretaries could venture support of projects off the beaten track, such as the aeronautical work of S. P. Langley, the third Secretary of the Smithsonian. The limited resources of the Institution, however, prevented it from acquiring a major role in the growth of science in America. The organization is a good example of institutional proclivities for drifting into the path of least resistance.

Bache's position as a scientific Pooh-Bah and the range of his agency were growing, but he wanted to be more than a gray eminence in other fields. He wanted to have some direct executive or administrative voice. It so happened that Henry's Smithsonian Institution was a better base for such an ambition than the Coast Survey. Perhaps Henry and Bache recommended each other for the wrong jobs. Bache, the empire builder, might have had greater scope at the Smithsonian; Henry might have expanded the weights and measures function of the Survey into a true physical science laboratory.

In the 1850's none of this was apparent. Henry and Bache were in the saddle. They had allies, and they controlled the principal scientific organizations in the nation.

At first, Henry and Bache probably expected great things from the American Association for the Advancement of Science (AAAS). In 1840 the geologists had formed the Association of American Geologists, which became the Association of American Geologists and Naturalists in 1842. Six years later, largely through the efforts of William C. Redfield, who served as its first president, the AAAS came into existence. At its meetings the Bache-Henry circle, under the title of Lazzaroni or scientific beggars, had their festive dinners. The AAAS gave promise of becoming a Lazzaroni preserve; Redfield (not a member) was succeeded by Henry, Bache, Agassiz, and Peirce. (The latter's letter of September 21, 1851, below, on the AAAS is a fine example of Peirce's rococo prose and characteristic ebullience—both delightful if taken in small doses.) But the AAAS was too all-inclusive and contained too many natural historians and amateur polymaths for the tastes of Bache and Peirce. Henry was much more tolerant and less eager for any grandiose academy of savants. In those happy days at the AAAS meetings before the Civil War were planted the seeds of the one great clash between Bache and Henry.

Henry to Gray

Confidential Princeton Dec. 12th 1846

My dear Dr.

I have been so much overwhelmed with business since I last saw you that I have not found leisure to give a moment's attention to the article I promised.

The die is cast I am sold for the present to Washington. I have accepted the appointment at the solicitation of some of the friends of Science with the hope of saving the generous bequest of Smithson from utter waste. I have forwarded a plan which if I am permitted to carry out will I am sure render the Institute of the highest importance to science of our country and the labours of every true working man of science among us.

If I find that I cannot succeed in carrying my plans and that the money is to be squandered on brick and mortar at Washington I shall resign and leave to others the honor of the perversion of a whole bequest.

I shall endeavor to stay proceedings at Washington and get time to elaborate more definitely my plans by conversation with scientific friends and otherwise. I must see Peirce and yourself.

I regret that the Regents have published a report of their plans because when once committed a politician can never change his course.

I start for Washington this morning. Since my appointment I have been overwhelmed with business and feel rather over worked.

I have made up my mind to the most disagreeable notoriety of newspaper praise and abuse. I have endeavored to be ignorant of what is said of me and shall continue to do so unless my honor is impeached.

Your friend
Joseph Henry

Gray Papers, Harvard University

Henry to Loomis

Princeton April 22nd 1847

My dear Sir

Your favour of April 12th came to Princeton while I was in Washington and I now begin this answer at almost the first moment of leisure I have had at my command since my return. I have nothing at present to suggest with reference to the plans of the memoir. So far as you have given it in your letter, it fully meets my views. I can give you no information until I see or hear from Mr. Espy as to the number and character of the observations made at the military posts. Mr. Espy's salary was struck from the appropriation bill at the last session of Congress and in consequence of this he intends to leave Washington next July.

I do not think that he can have any cause to be displeased with our proceedings. You will of course give him due credit for his labours and he will be invited to furnish for publication in the Smithsonian contributions to knowledge the results of any researches he may have as yet not given to the world. I would however be glad if it was in my power to do something in the way of attempting to restore him to his former position. I consider him a man of most excellent character who has laboured industriously and successfully in the cause of science and who in a country of so much wealth as ours should not thus be deprived of the pittance to which a few months before he was thought entitled. He has continued to receive the reports from the several government stations but has not published any results that I have heard of since 1843.

The Secretary of the Treasury at my suggestion appended to his order for a geological survey of the new Territories, directions for a set of magnetic observations on the dip-intensity. The condition was that I should give the instructions and purchase the instruments, the results to be given to the Smithsonian for publication.

I found by enquiry at the land office that all the surveys of the public lands are now made with an instrument called a solar compass which gives the meridian by means of an image of the sun, the declination being known, to within about a quarter of a degree and perhaps less. Also the surveyor has been directed in all cases, on each line to note the deviation of the magnetic needle and in this way considerable material has been furnished for perfecting the variation chart of our country. Would not the plan of procuring a good map plate of the United States and having a number of copies made of it by the electrotype process be of interest? On one of the plates the magnetic lines being declinated, on another the thermal, on a third the geology etc., and thus in time forming the elements of a physical atlas of our country.[14]

I think of visiting New York in the course of a week or two and I will then give you a full account of all the proceedings relative to the Smithsonian. In the meantime I beg to assure you that I remain as ever

Truly yours
Joseph Henry

Loomis Papers, Yale University

Henry to Gray

Confidential Washington January 10th 1848

My dear Dr.

I presume you have been expecting a letter from me for several weeks past, and I regret that I have not been able to give you any account of the prospects of the Institution before this time. Indeed I am still unable to say what a day may bring forth. The Board continued in daily session for three weeks and since their adjournment I have been much occupied in preparing the Report and proceedings for publication.

The Report from the Amer. Acady. came in good time and served

[14] A good example of Henry's constructive ideas in geophysics. Probably nothing came of this because of the rivalry between the civilian agencies (then the Treasury—later the Interior Department) and the Army Engineers.

a good purpose. I regret however that it gave me the whole credit of the programme since it contained all the suggestions, so far as I thought them valuable, of all the persons with whom I consulted on the subject; and because by giving me the credit for the whole, offense would be given were the article published. The programme was provisionally adopted in full with a few unimportant additions and corrections made by the committee of organization to whom the article had to be submitted and I was charged with carrying it into operation so far as the appropriation of funds for the purpose would permit.

When I came on to Washington I found Owen busily engaged in devising a plan to increase the income of the Institution so as to cover the odium of the expenditure on so large a building. For this purpose he proposed that the operations of the Institution should be limited to an expenditure of 15 thousand dollars annually instead of 30 thousand until the end of 4 years from next March or until the building shall be completed: the other 15 thousand with its interest to be added indirectly to the principal and thus to make the annual income ever after 40 thousand dollars instead of 30. I at first gave but little attention to this matter. I was however surprised to find that Owen[15] had brought over every member of the Board in Washington to his scheme and when I found that it met the approbation of Bache I was induced to look upon it with some favor and at length to desire its adoption, at least in part, for I wished to encourage the spirit of economy which now appeared to actuate the measures of the Board and because I thought if Owen was once out of the Board the whole could afterwards be adjusted. There was however one stipulation which I made namely that Mr. Choate[16] should have an opportunity to be present and object to the Resolution if he were so disposed. For this purpose I sent him two telegraphic messages and one to Mr. Jewett.[17] The latter gentleman came on and by means of another message induced Mr. Choate to start for Washington. Mr. Jewett was very anxious to be immediately engaged in the duties of his office though by the terms of his appointment he was not to begin

[15] Robert Dale Owen (1802–77), son of Robert Owen, then a member of Congress. Owen designed the turreted, towered Smithsonian Building believing its architecture in harmony with the American spirit. It was a white elephant disliked by Henry as a drain on the funds. Owen also pushed the popularization of science rather than original investigations.

[16] Rufus Choate (1799–1859), a Massachusetts lawyer and a strong advocate of a large library.

[17] Charles Coffin Jewett (1816–68), the Assistant Secretary and Librarian of the Smithsonian. A great librarian, Jewett was later dismissed by Henry to avoid the conversion of the Institution to a large library.

his services until the Building was in a fit state to receive the Books. Mr. Choate at first objected to the Resolution and insisted that the purchases of books should be commenced immediately. He however I think came to regard the proposition more favorably, and on condition that Mr. Jewett should be immediately employed in the way of preparing catalogues and making arangements for the purchase of books at the proper time the Resolution passed. During the present year I shall have about 7000 dollars to expend in the way of memoirs, experiments apparatus, etc. and as I am not anxious to push the operations too rapidly, I am, for this year, content with this sum. If however all things go on well we shall require a larger sum the next year which must be drawn from the interest and the desired result (the increase of the fund produced by the extension of the time of putting up the less essential parts of the building).

I had no idea at first that Mr. Owen would have any chances of getting into the Board after his time expired, Mr. Rush has signified a desire to remain a Regent during his absence and there was therefore apparently no vacancy. I was however surprised by the proposition from one of Owen's friends that Mr. R. should be made an honorary member and Mr. O. elected in his place. Finding that Owen would probably get in I had a free conversation with him and insisted that he should resign his positions in the executive and organizing committees and confine his whole attention to the building. To this he agreed. The Board afterward on motion of Gen. [William J.] Hough [of the House of Representatives] recommended Mr. O. as the successor of Mr. Rush. His appointment however has not yet taken place. He will probably be nominated in the Senate and the nomination sent to the House for concurrence. I know not what will be the result.

At the beginning of the sessions of the Board I presented my report which should have gone through the hands of the committee on organization in accordance with the resolution of the Board under which I acted. I showed it to Bache and Hilliard and then presented it to the Board. It was well received and entered at large on the minutes. A resolution was adopted directing the secretary to report annually the conditions and operations of the Institution so that in future I shall not be controlled by a committee.

The executive committee now consists of Seaton,[18] Bache, and Peirce, and with the two latter back of me I think all things will go well during the present year. I hope to be able to receive Peirce's paper in the course of two months. Ethnological Memoir is not yet

[18] William W. Seaton (1785–1866), a journalist, then Mayor of Washington.

published; it could not be completed until after the meeting of the Board because I had but 1000 dollars to expend on the article. I start tomorrow for New-York to make the final arrangements for the printing of the Memoir of Squier and Davis[19] the wood cuts and plates are nearly finished.

The attention which Squier has received from some of the great men in Boston and New York has nearly turned his head and caused him to give me considerable trouble.

I will write to Peirce relative to his paper. What I have given you is confidential.

As ever your friend
Joseph Henry

There is a proposition in the House to look into the affairs of the Institution by the appointment of a committee. This I think will pass.

Gray Papers, Harvard University

Copy of a letter from Baron Alexander von Humboldt to Dr. J. G. Flügel

Leipsic.[20] (Dec. 22, 1849)

I hasten, my respected Doctor, to express to you my best thanks, though only in a few lines (being engaged since yesterday in all the horrors of a moving from Potsdam to Berlin) for the very interesting Journey of Emory and the Notices of Gilliss, referring to the ascertainment of the Parallax in Chili, and the astronomical Longitude of Washington. For Major Emory's[21] work, I had specially written to New York.

For an [Antidiluvian?][22] who, like myself, is attached with his whole Soul to the New World, through the coloring of opinions, and the Knowledge of the beautiful (a country spotted only by its legalizing of Slavery) it is renovating and pleasing to follow and quick and proud Expansion of Scientific sentiments in the United States, and to have to acknowledge how much the Government participates in an expedition to Chili of three years, because a Professor of Marburg is wishing for it, and nobody listens to him in Europe!!!

[19] The memoir was on the mound builders. Ephraim George Squier (1821–88) and Edward Hamilton Davis (1811–88) both had distinguished careers as archaeologists.

[20] U. S. Consul at Leipzig.

[21] W. H. Emory (1811–83) an Army Engineer.

[22] Question mark and brackets in original.

We are indebted for very valuable hypsometric, astronomic, botanic and geognostic Works to Charles Fremont, Emory, Wislizinus, Lieut. Abert,[23] and Bache of the splendid Coast Survey, the Circumnavigations of Charles Wilkes . . .

Coast and Geodetic Survey Records, National Archives

Peirce to Bache

Harvard 21 Sept. 1851

In what tone shall I address him? Shall I assume the lofty and contemptuous air of justly offended dignity; combining presidential sublimity with scientific sublimation? or shall I be the Jupiter Titans and in the tornadic wrath crack my hurricanes over his devoted head?

Earth-born Geodetic! Slave of the screwworms with which you strive to measure earth and sky! Crawling to the hill-tops in hopes to find some firmamental flaw, through which you may peep into the secrets of the fair stars! In vain do you fabricate your spy glass key-holes. Vain geodetic gossip! Mighty tidy lunatic! Down to your brothers valets!??? Boreal bears! Breathe the northern wind over him! Dash the polar axis through his brazen circles! and heap icebergs upon his tents! With what epithet shall I dishonor his prowd name? O! once lived professor! how dearly loved! I am, I am sick! O heart! atone me or I fall! There, I am myself again, and now how far you are behind the age to foster your indignation when Foster[24] has forgotten his! His last letter commenced "My Dear Sir", and he now remembers that I once told him that I was very apt in fits of absent mindedness to pass my nearest friends unnoticed. You may now then consider yourself as uniced and melt back again to pristine familiarity without danger of offending our Peirce prowd president. The constitution declares that the standing committee shall consist of the officers of the preceding year and six members to be chosen by ballot. What form of ballot is here intended? I say that which the association may elect. In my opinion it should as far as practicable be an open ballot, that is, each vote should have written upon its

[23] John Charles Frémont (1813–90), the noted explorer, later entered politics, unsuccessfully running for President in 1856. Frederick Adolph Wislizinus (1810–89), a German physician who settled in the United States, and wrote accounts of his travel in the West. James William Abert (1820–97) was an Army Engineer engaged in western surveys.

[24] Probably the geologist John Wells Foster (1815–1873), later (1869–70) President of AAAS.

back the name of the member by whom it is deposited; and no member should be allowed to vote who has not paid his fees. Should not absent members vote? What is your opinion upon these matters? What regard is to be had in the selection of a president to his having previously presided? Have we not silently acted upon the principle that to have enjoyed this honor once is often enough? Enough, at least, until the list of proper candidates is exhausted? In this case, are not the presidents of the geological association, former presidents of this association?

Walker has returned in very fine spirits, and seems better than he has for a very long time. [Benjamin Apthorp] Gould[25] has had the official offer of Professorship at Göttingen, with a promise of Gauss's influence to obtain the reversion of the directorship of the observatory. We have advised him, by all means to accept the offer. The Prussian times of the solar eclipse tell the same story with the American ones regarding the new tables. They confirm Longstreth[26] but I am doubtful whether the new correction of parallax satisfies them so well as the old parallax did. It is certain that there is no indication of the proposed augmentations of solar and lunar semi-diameters.

Mrs. Peirce sends her kindest greetings to her faithful nurse, whose prolonged absence is audibly attested by the continual weeping of the deserted child. Give her and my own warmest remembrances to Mrs. Bache. I hope that you have confessed your flirtation to her. I thank my stars and Natural Birds Eye that I have none to confess. Good bye my very own and true friend, first in science, first in sagacity and first in the hearts of your friends, of whom none is more devoted than myself.

Benjamin Peirce

Peirce Papers, Harvard Archives

[25] Gould (1824–96) was apparently the first American offered a professorship in a European university. He refused; his subsequent career never quite lived up to his early promise. He founded the *Astronomical Journal* in 1849 and, except for a four year interval, he was with the Coast Survey until 1870, when Peirce (who had succeeded Bache) forced him out. In 1855–59, he headed the Dudley Observatory in Albany until practically run out of town by the outraged citizenry. Apparently his personality is a partial explanation of the lack of fulfillment of early promise. From 1870 to 1885 Gould was in Cordoba, Argentina, where he produced valuable results observing the less known heavens of the southern hemisphere. In the 1850's and 1860's he was still in the good graces of the circle—with curious results for the history of the National Academy of Sciences.

[26] Meirs Fisher Longstreth (1819–91), a Quaker physician in Philadelphia who maintained a private observatory.

EVOLUTION: FROM NATURAL HISTORY TO BIOLOGY

Natural history, not geophysics, was the most widely pursued scientific activity in nineteenth-century America. To the geophysicists, the men measuring the earth and the heavens, much of natural history must have seemed a mere matter of describing little details and lacking in the exactitude of the physical sciences. Natural history was non-quantitative and even less committed to experimentation than geophysics. While the physical sciences already displayed deepening fissures between amateurs and professionals, natural history still was hospitable to the amateur and the polymath.

The nineteenth century was one of the great ages in science, and it is futile to argue whether the physical or the biological sciences advanced most. Certainly, the developments in the latter are a notable chapter of intellectual history. The best known of these is the work of Darwin and his associates. Darwinian evolution summed up much of the accumulated knowledge of life, cleared away extraneous intellectual underbrush, and stimulated many research areas. Not as well known, but probably as significant as the concept of natural selection were the growth of physiology, the development of microscopy, and the rise of biochemistry. Natural history, an ancient brew, changed into biology, a modern elixir, as a result.

Few Americans played notable roles in this change. They were conspicuously scarce in the ranks of the microscopists, physiologists, and biochemists prior to 1859, the year of the publication of *The Origin of the Species*. The very few who played any role of consequence in these fields were odd exceptions to the prevailing American scientific patterns. An example is Dr. William Beaumont (1785–1853), an Army surgeon, who studied the physiology of digestion using a living subject, a man who had acquired a hole in the abdomen in an accident. Beaumont's work was first-rate and influential, but no Beaumontian school of American physiologists appeared in his wake.

Americans contributed to the Darwinian theory in two interrelated ways. The first was often unintentional: flora, fauna, and rocks collected by American scientists sometimes provided data for Darwin. Geologists, a very active scientific group in nineteenth-century America, were particularly important because of the fossil remains they brought to light. The second contribution was made by a small number of Americans who became correspondents of Darwin's, providing him with information and criticism. The most important of these was Asa Gray, the botanist.

Geology and Paleontology

By 1860 American work in geology was both considerable and sophisticated. The efforts of individual scientists were aided and supplemented by the activities of the federal government and the states. Although the interest of a state in a geological survey was usually utilitarian—to discover mineral wealth—and the Army explorations in the West were primarily for mapping, the results of such activities often went beyond narrow utilitarian bounds. Enlightened officials and energetic scientists found ways of expanding the surveys and explorations into general natural history research projects. Economy-minded and utilitarian legislators might fume and on occasion successfully move to cut off funds, but the over-all trend was for an increase in support of geology and natural history. Mountains of specimens were gathered, and research expanded. By the end of the century American geology was at least equal to that of any other country both in terms of quantity and quality.

In retrospect it appears that the geology of this era both suffered and profited from an ambiguous role, being at the same time part of natural history and of natural philosophy. In its best exemplars, such as the Englishman Sir Charles Lyell or the American James Dwight Dana, geology was concerned both with the physics and chemistry of the materials composing the planet and with the discovery and classification of extinct forms of life. There was a tendency to study the rocks in a Linnaean spirit, as a problem in taxonomy. At the same time knowledge from physics, chemistry, astronomy, and geophysics were applied to the problem of the dating of geologic events. From the study of fossils it was also possible to uncover the sequence and timing of such occurrences. The two sets of data, fossil remains and the physical factors, served as checks on one another, stimulating further research when they did not agree. The history of geology in the past century can be partly interpreted as a resistance on the part of the discipline to polarization into natural history by way of paleontology and into a minor branch of the physical sciences, such as mineralogy. As a result, a distinct scientific field developed.

In the letters that follow we first see the normal business of specimen collecting and exchange involving James Hall and the Sillimans. Next comes examples of European interest in the American findings—Sir Charles Lyell and Hall—and a violent but not too untypical reaction of one American scientist (Hall) to supposed European disregard for professional niceties. The last letters, mostly correspondence of the great paleontologist Joseph Leidy with F. V. Hayden, the geologist, are good examples of the routine and not-so-routine trials of getting parties to the West, getting specimens home, and financing publications and new surveys. No sensational theory resulted from Hayden's field work, but it was as much a part of the life of science as the theorizing of Darwin. Ultimately the labors of Hayden, Leidy, and their kind would provide strong evidence for evolution.

James Hall [1] to B. Silliman, Sr.

Cohoctan Steuben Co. July 22nd 1839

Dear Sir,

I have just received the July No. of the Am. Journal of Science, it having been brought me by a friend from Albany. I observe that you advertize for Geological specimens, fossils etc. of good quality and well preserved. If you want those which the state of New York will furnish I think I can send you a valuable collection. The rocks and fossils of this state will furnish an interesting and instructive suite, being the exact counterpart of the groups of rocks of the Silurean system as described by Mr. Murchison. We have besides this advantage, that our rocks are better developed, have been subjected to less disturbance and contain a much greater number of fossils. I think we shall find nearly every one figured by Mr. M., besides many others. I have lately found the new genus of Tribolite which he figures, the Bamastis, and in some examinations within a few weeks have discovered the remains of fossil fish in the Old red sandstone, which is an interesting and important member of the series, though it has usually been overlooked. And I am not aware that the remains which I speak of have been before noticed. They are the scales, jaw and fin of a large fish, perhaps the Gyrolepis—the scales are an inch or more in length and nearly as broad. I at first supposed them to be of the Cephalaspis, but on examination find they are not. While engaged in the Geological survey I do not like to engage to furnish rocks and fossils very extensively as it might be though[t] to interfere with my duties to the state, but I have many on hand which I have been collecting for several years past. It is also

[1] Hall (1811–98) served New York State as geologist and paleontologist for almost sixty-three years. He was especially notable in the later capacity, specializing in invertebrate fossils as well as the more popular vertebrates. Hall's work was important because his series of rocks and fossils served as a basic standard. His name is also known for a theory of mountain building. An irascible, great figure in American geology.

my intention so soon as the survey is completed to engage extensively in collecting the rocks, fossils and minerals to illustrate the Geology of New-York in the most perfect manner. For this enterprise I shall be glad of your contenance and aid, but at present I do not wish to make the matter public.

Will you have the goodness to state to me the number of specimens wanted, the price you intend to pay, and also if a complete suite of the Silurean rocks and fossils will be desirable, also what others. I can without much difficulty furnish the fossils of the Cretaceous group of New Jersey, and if the price you will pay will warrant it I shall be glad to supply those of the Tertiary. The Primitive rocks of this state as well as minerals can be readily supplied, principally from what I now have on hand. The rocks and fossils of the Silurean system cannot be collected properly except by someone acquainted with the whole series from minute examination. For my ability to do this I will refer you to Mr. Conrad and Mr. Vanuxem, and also to Dr. Emmons and Mr. Mather.[2]

I shall not return to Albany till November, my address till that time is Gorham Ontario Co.

I forwarded by gentleman a copy of the Geological report to Yourself and one to the Yale Nat. Hist. Soc. Have you received it.

With great respect
Your obedient servant
James Hall

Hall to B. Silliman, Jr.[3]

Albany January 15th 1840

Dear Sir,

I received your obliging letter of Dec. 11th, which I should have answered before this time but have been very constantly engaged in preparing my report for the Legislature. I am greatly obliged to your Father and yourself for the favorable manner in which you have your desire to forward my object in regard to the geological collections. I will prepare a notice for the April Number.

I have been writing a few pages of a kind of review with short extracts from Murchison's work, and illustrating by comparisons

[2] Timothy Conrad (1803–77); Lardner Vanuxen (1792–1848); Ebenezer Emmons (1799–1863); and W. W. Mather (1804–59)—all amateur geologists.
[3] The younger Silliman (1816–85), a chemist, was also an editor of the family periodical.

with our rocks. Perhaps I may be able to make it interesting and in some degree useful, and if agreeable to you will send it for publication in the Journal of April.

Dr. T. Romeyn Beck is the proper person to furnish notices of the proceedings of the [Albany] Institute. He has been absent for several weeks but will return in a few days. If I can be of service to you in this matter I shall, do so with pleasure.

Your Nat. Hist. Society shall be remembered in the Spring. I had laid aside a small collection a year since but did not complete: however you will be no loser by the delay. I will gladly embrace your suggestion of opening a correspondence with the Museum at Heidelberg. I will state to Prof. Bronn[4] that you suggested it, and I presume that may be a sufficient guarantee that my collections are worthy his notice.

Can you inform me how I may send a letter to Prof. Agassiz? I wish to send him a drawing of the scale of a fossil fish which I cannot find described in any work I have met with. It may possibly be of the Genus Megalicthys. I have seen no description of this Genus. I have thought of proposing Megalepis from the large scales.

I will endeavour to send to your Society a part of what I intend for my geological notes of New York. I cannot be prepared with the whole without devoting at least two years more labor to the collection of rocks, minerals and fossils.

With great respect
Your obedient servant
James Hall

Silliman Papers, Yale University

Torrey to Hall

New York July 21st 1841

Dear Sir

I have just received a letter from Mr. Lyell[5] announcing particularly, what you probably had heard intimated before, that he proposes spending some months in this country in exploring some of the more interesting geological regions. He was to embark at Liverpool in the Acadia, on the 20th of July and hopes to reach Boston by the end

[4] Heinrich Georg Bronn (1800–62), the German geologist who was one of Agassiz' teachers.

[5] Sir Charles Lyell (1797–1875), perhaps the greatest geologist of the last century.

of the first week in August. He wishes then to devote 8-10 weeks, to exploring the geology of the State of New York, and the country about the Falls of Niagara and Lake Ontario. He wishes to have your company and to know where you may be found soon after his arrival. He desires me to inform you that he has just returned from a geological tour in the border Country of England and Wales where he has been examining the older or Silurian strata with a view of comparing them with those of the United States. He has also made a collection of fossils which he will bring with him to Boston.

Please communicate to me your wishes with regard to the letter of Mr. Lyell, or send me a letter for him and I will take measures to have it delivered to him on his arrival at Boston.

Yours truly
J. Torrey

Merrill Collection, Library of Congress

Hall to Benjamin Silliman, Sr.

Albany March 28th 1842

Dear Sir

When in Boston[6] I intended to send you these copies of a part of Mr. Lyell's illustrations long before this time but Dr. Emmons was intending to lecture in Williamstown and wanted them. I have had copies made of all essential ones. I mean all that are not found in his books or in other treatises in use in the country. You will perceive that Niagara Falls is somewhat altered from Mr. Bakewell's[7] sketch but essentially the same, this was done after we had visited the falls together. The Coral island is a better drawing than the large one made by Mr. Emmons, which I could not for the reason mentioned send you. I hope some of these may be of service to you, if not I shall put you to expense for the carriage. These can be returned at your pleasure. I have no use for them for six months.

The points of inquiry regarding some things in Mr. Lyell's lectures are not as numerous as I supposed. One in regard to the number of species of plants of the Carboniferous period, he gave 300 as of that period alone, though it is greater than any list I have seen.

The other regards the formation of coralline limestone at Niagara and afterwards a sinking down and a deposit of 1000 feet or more

[6] For the meeting of the Association of American Geologists.

[7] Either Robert Bakewell (1768–1843), an English geologist, or his son of the same name, who settled in New Haven. Both may have written on Niagara.

of mud with no coral, and this succeeded by coralline limestone at Black Rock.

Such is the fact however relating to these rocks, as you will perceive by referring to a little wood cut in my [article] in the last No. of the Boston Journal of Natural History.

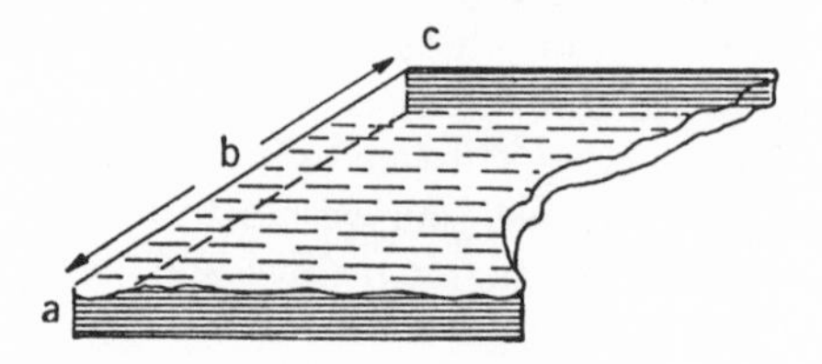

a. represents Niagara Falls, b. a space of fifteen or eighteen miles occupied by an agillaceous [?] deposit containing gypsum, c. a coralline limestone at Black Rock. The deposit b. cannot be much less than 1000 feet in thickness. I believe there are no other points of inquiry. I perceive you made some corrections for me for which I thank you, though I regret that I was compelled to send you such a manuscript.

Should you have any work on fossils which you can lend me for a time during the summer it will greatly oblige me. You mentioned some copies of Von Buch's[8] works which I have not seen and they are not in our State library and I presume not in Albany.

I learned a few days since that Mr. Lyell had made arrangements with Wiley and Putnam of N.Y. to publish an edition of his Elements with notes and additions on American Geology. You may well suppose that I was amazed, and can it be possible that Mr. Lyell will take this course after all his repeated declarations that he should publish nothing till after the appearance of our Reports here?

If every other one has the same feelings that I have on the subject there would be a strong expression from the Association at its meeting, denouncing what I consider piracy in its worst form. Mr. Lyell has written me twice since that time and does not mention it. I learned moreover that he is to remain another year in this country to prepare this new matter. I say this to you in confidence. Mr. Lyell has always treated me with more attention than I could expect, and if I wrong him here I am sorry, for it. But after having spent my time and money to explain to him the structure of the

[8] Baron Christian Leopold von Buch (1774–1853), the best known German geologist of that era.

rocks of N.Y., in all which I kept back nothing I think I have a right to feel. By a few weeks in this way he has learned what has cost us years of labor and which he is now to palm upon the Gullible American public as his own. Already the newspapers are lauding him in advance. And if this be the case all the doings at our meeting will go the same way; every fact in relation to American Geology will be seized for his own use. I am sorry that I am compelled to think thus of Mr. Lyell.[9]

I am desirous to know whether anything of importance will transpire on Monday 25th the first day of the Meeting in Boston: I did not intend to leave here till Monday morning and I presume Dr. Emmons will not, and Mr. Vanuxem will also be here to come the same way.

I remain with high regard
Your obedient servant
James Hall

Silliman Papers, Yale University

Ferdinand V. Hayden[10] to Hall

St. Louis May 10th 1853

Prof. Hall,
Dear Sir,

Dr. Evans has just arrived on his way to Oregon. He will first go to the "Mauvaises Terres", make a collection, send them down the River, and then proceed on his expedition. I spent most of last evening with him.

Father DeSmet[11] has gone to Boise, started 25th April. Rev. Mr. Murphy will be absent 2 months. I cannot see either of them. I have been very busy getting information as to the best course to pursue.

From the best information I can obtain I am satisfied that the land route is far preferable. That is, go by steam boat to St. Josephs fit out there and go to Fort Laramie and thence to the "Badlands."

[9] Hall's feeling became public knowledge, many condemning him, at first for aiding in a foreigner's intrusion into American geology. Later, Hall was taxed for an injustice to Lyell. Lyell disregarded the tempest, gave Hall due credit, and continued to associate with the Albany geologist.

[10] Hayden (1829–87) was a physician who became a leading geologist. After the Civil War he directed one of the four great surveys from which sprang the U.S. Geological Survey.

[11] Pierre-Jean DeSmet, S.J., (1829–87), a missionary to the Indians.

I spent some time with a Mr. Campbell been a merchant in this City, who has spent many years in these regions, and owns a trading post, high up above the Yellow Stone and a great trader. He very strongly recommends to me the land routes as far cheaper, more pleasant, and more certain of success. He has promised to give me letters to different men of his acquaintance who will assist me on my way. But I can make no decision until Mr. Meek[12] arrives and hear something from you. I have not as yet heard a word and it is a source of a good deal of perplexity to me I will explain the reason why I prefer the land route as far as I understand it now. If we go up in the Boat to Fort Pierre, we cannot make the Collection and return in three months for less than a thousand dollars. Mr. Lesley one of the partners told me, if we go the land route the same amount of money will keep Mr. Meek and myself 6 months, bring him back and continue my labors a year longer. If this is true (and it appears so to me) it is as such worth considering. If we go the Land Route, we go from here to St. Josephs by Steamboat carrying our provisions Baggage etc. a good light California Wagon. At St. Josephs we get two good mules, put our effects in to our wagon and go with the numerous emigrant trains which constantly fill the road. After leaving St. Josephs our expenses will be nothing, the travel through a fair country in which a thousand things of interest may be found, and as safely as on the Boat, from St. Josephs to "Badlands" 750 miles, at 25 miles per day, 30 days. Mr. Evans[13] thinks we can get there as soon or sooner than to ride from Fort Pierre. When we arrive there we have independently of any one all our fixings, two good mules and a good wagon. At Fort Laramie we can get plenty of help cheaper and better than at Fort Pierre. Many of the Soldiers or even Officers will like to relieve the tediousness of situation in that way. If not there are a plenty of [voyageurs] that may be obtained. At Fort Laramie I am told, we can live much cheaper, and in all probability cost me nothing to stay through the military. Our *"treasures"* we can transport to Fort Pierre and have them sent down by boat or they can be sent from Fort Laramie in the Quartermasters wagons which always return empty. Mr. Evans thinks that will be the better way

[12] Fielding B. Meek (1817–78), a collaborator of Hayden's and a leading geologist.

[13] John Evans (1812–61), a geologist, especially noted for work in the Pacific Northwest.

and that he will probably send his collection down in that way. If we go up by Boat, the company will furnish our supplies, charge us 50 dollars each for our passage up, charge us for our freight. When we get there they will furnish us men at from 2 to 3 dollars per day and an old cart that will break down the first 100 miles, at one dollar per day, 2 mules that have made out to flinch through a cold winter with a small remnant of life left for one dollar per day each. They will expect to furnish us food for ourselves and them, at about three times its real value and poor at that. Every thing we get, we must pay at least their prices or they are willing to afford us all the facilities in their power. But how do they do it? Why, for any thing that they do or furnish, we must pay them an enormous price and make them as little trouble as possible.

It is certain that the Germans are going on the Boat. They have made all their arrangements, paid their passage, and secured the Company assistance. This they did some months ago, Mr. Evans did the same, though he says that if he had not been delayed so long in Washington he should have taken the Land Route. Now we come in for our aid after them. They select their mules, men, 8 carts and then we will stand our chance.

As to being behind hand, there is a chance for us, if we are diligent to get there even before they do. And if we should not, Mr. Evans says, that after we have all labored two years, those that come after us will have just as good a chance and perhaps better even. But a small portion of the Bad Lands has been explored, and, he says, there is no reason why there should not be other Localities as good yet to be found. Mr. Evans goes to Mauvaise terres by direction of Prof. [Spencer F.] Baird which may account for his not noticing your requests. I wish you would consider what I have written and if you think best, telegraph to Dr. Englemann.[14] Otherwise when Mr. Meek arrives we will do what seems best, I prefer the land route with all the risks, and I prefer to go independent if I have to sit up all night and watch the mules and sleep in the wagon days, rather than let the Fur Company speculate out of me.

I have been busy getting instructions and making arrangements as far as I could, it is much to be regretted that Mr. Meek would not have started at least a week earlier. However I will have matters so that a day or two will fit us out to go either route as it may seem

[14] George Engelmann, a German immigrant to St. Louis. A physician, Engelmann was greatly interested in botany.

best. If we go by land there must be some different arrangement in regard to funds. It will cost about 350 dollars to fit out and get started from St Josephs, after that, the expense will be trifling, until we arrive at Fort Laramie. After we get through with our team and things, they can be sold for at least one half their first cost. Dr. Englemann has just inquired the cost of a good wagon. He finds a man who will make a good wagon, fit it all out with cover etc, warrant it, for 70 dollars. He needs but a weeks notice to have it ready. In what way we wish we need a letter from the Secretary of War to the quartermaster like that one in that little book of instructions of Bairds and sent on immediately. Dr. Englemann will see that it is sent to Fort Laramie. I have written as I think at present, perhaps when Mr. Meek comes, things will look different. I shall be satisfied either way. Dr. Englemann wishes to be remembered to you and will write when I leave. I have not written very plainly for my pen is not as good as a stick. I shall write again soon.

truly yours,
F. V. Hayden,

Sir Charles Lyell to Hall

Tremont Hotel Boston, 8th June 1853

My dear Hall

We find on our arrival here that the exhibition of New York is put off and will not open till the 15th July. I hope before that time that I may be able to visit Albany if you were to be there, but I am not sure till all the Commissioners have mustered and determined our course of proceeding under this unexpected turn of affair. For up to the time of our departure we were assured that the 1st or at latest the 15th June would be the opening day.

Please to write to me at once at Boston if at home when this reaches you. I should like much to have a few days geology with you on the question now so much controversed, the passage of the sedimentary into the metamorphic and the denial of Darwin and Sharpe[15] that in gneiss we can see any remains of what was once stratification.

But I am not sure of being free even for a week, tho' most probably I shall be. Do you go to Cleveland Ohio? And when is the

[15] Daniel Sharpe (1806–56), an English geologist.

meeting?[16] Have you any geological news? Dr. Leidy's[17] making out two cretaceous in the New Jersey green sand or chalk, which he calls priscodelphinus interested me much.

I send you my paper on the first American carboniferous reptili, and the first pulmoniferous mollusk of the coal and I hope you will admit I was not too sanguine about them last autumn.

yours truly
Charles Lyell

Lyell to Hall

11 Harley St. London 24th Oct. 1853

Dear Hall

I had hoped ere this to hear from you to report progress and shall be too happy if Prof. Wilson[18] who is expected here by the end of the week shall be the bearer of the Report by you. I am the more anxious as I have every expectation of starting from London about the 24th of next month, (a fortnight earlier than my former plans) on my way to Teneriffe and the Canary Islands.

This acceleration of my voyage has been necessitated by the quarantine regulations which Spain is believed to have imposed everywhere and which are already in force against all England because one or two ports in the north are infected with cholera and some spots in London, tho' the weekly mortality here is below the average. I shall have to go via [two words unclear] coast of Spain and Cadiz.

I write to day in the hope of hearing an answer from you which would reach me if you send by return of post, before I leave town. Always supposing you have not written fully and possibly sent off your dispatch by Prof. Wilson.

I believe I told you in my letter of Sept. 26 of the delay in my letter reaching Wilson which twice crossed the Atlantic. In the same letter I gave you till the first week in Nov. to get the report into my hands. Shall I get it by that time I shall be able to see it thro' before I leave. If it had to wait till I return in the spring it would be awkward for me and I am sure you would be equally

[16] Meeting of the American Association for the Advancement of Science.

[17] Joseph Leidy (1823–91), probably the greatest naturalist in America in the last century. Although best known for his paleontology, Leidy worked in many fields. He was trained as a physician and taught anatomy at the University of Pennsylvania.

[18] Prof. John Wilson (1812–88), a British agriculturalist, a Royal Commissioner to the international fair in New York in 1853.

concerned, altho' circumstances may have impeded you not under your control. Certainly we cannot perform impossibilities and if part of an exhibition is not open we cannot report upon it. They (government here) always knew I shall go to the Canaries.

Was it Benjamin Silliman who was ill or his father?

You see tha[t] Leidy doubts now about *both* specimens of his Priscodelphian being *cretaceous*. I refuted one of them with Conrad's help when in New Jersey.

Wetherils specimen of a mammals tooth appears to me to stand on good grounds and it was in the lowest New Jersey beds.

If you have any news bearing on "progressive development"[19] do not forget to send them. I have received the money which I think will procure a second pupa from the Nova Scotia coal.

With rememberance to Mrs. Hall believe me ever truly yours

Charles Lyell

Merrill Collection, Library of Congress

Leidy to Hayden

Philadelphia December 6, 1854

Mr. F. V. Hayden
Dear Sir,

I am pleased to hear again from you and will be delighted to see the interesting saurian remains of which you speak. For whom are you now collecting? How are you especially engaged?

I am informed that about 32 miles this side of Fort Laramie near the road, there is a remarkable locality for fossil bones. They are stated to be embedded in a sandstone nearly as hard as granite. It forms the base on which the clay of the Mauvais Terres rests. Beneath the sandstone is a hard limestone containing fresh water shells.

Nothing has yet been published in relation to Nebraska; neither by Prof. Hall nor Dr. Evans. The work I have undertaken is slowly going on. In your collection, besides the two specimens characteristic of new species from Bijm Hill, there is the head of a giant weasel belonging to a new genus; to which I have given the name of *Dinictia felina*. It had long upper canines flattened, and sabre shaped. One of the two specimens from Bijm Hill indicates a new species of the solipedia the other of the ruminants. I think the last mentioned locality one well worthy of examination.

[19] i.e., evolution.

I shall be glad to hear from you whenever you feel disposed to write.

With respect,
Joseph Leidy

P.S. If I can be of any service to you, or can in any way promote your objects let me know it. If you wish to make a sale of any of your specimens, for further explorations, I will see what may be done by subscription in our Academy.

With respect,
J.L.

Leidy to Hayden

Philadelphia July 8th 1855

My dear Dr. Hayden,

I am delighted once more to hear from you and of your success in making collections in natural history.

In a late letter of yours, which I did not answer from uncertainty how to direct it, you observe you had made an agreement with Col. V. to be permitted to examine your collections on arriving at St. Louis. You do not mention the gentleman's name in full, but from what you observe I suppose you have been collecting for him.

You also spoke of a very liberal undertaking on the part of Mr. Gilpin, but I am doubtful of the results of the enterprize, for which I feel sorry as it would prove in the highest degree of benefit to science.

I would like to examine and describe the new vertebrate remains you have discovered and would be glad to purchase a set for the Academy from which I would try and obtain the highest price on the value.

If you have collected for another person, and obtain for me permission to describe the vertebrate remains in Dr. Evans forthcoming report I will pay the freight, give to you twenty-five copies of the complete memoir and $200. to be received from Government in recompense for preparing the latter.

Or, if the gentleman for whom you collect and yourself prefer it, if your new discoveries are sufficiently large I would prepare a distinct work with such a title as "Description of extinct vertebrata from Missouri, collected by Dr. F. V. Hayden in the year 1854, 5 for Col. V.", to be published by the Smithsonian Institution, and

Col. V. and yourself would receive one half the copies given by the Institution to the Author, the whole number usually being 100. In the latter case the things should be sent through the Smithsonian Institution for me, with a letter to Prof. Henry, stating their destination and the object, by which we would save the freights, and if the new material is large I would give to you a proportional amount of money as a personal contribution to science. Indeed I wish to do all that is in my power.

Dr. Evans' collection from Nebraska weighed nearly two tons, of which about 800 lbs consist of uncharacteristic fragments, eventually to be thrown away as utterly worthless. . . .

The Smithsonian Institution has just published for me "A memoir on the Extinct Sloth-tribe of North America quarto, 58 pp. 16 plates." I wish I could send you a copy, but suppose I must wait until your return.

If there are any persons you meet with especially interested in the Nebraska fauna and you would like to give them a copy of "The Ancient Fauna of Nebraska"[20] I will place some at your service.

Lastly as a tribute of respect to your zeal I have proposed you as a correspondent to our Academy and hope the next time of writing to announce your election.

With respect I am at your service; and hope you will let me hear from you shortly.

Joseph Leidy

Leidy to Hayden

February 25, 1856

Dr. Hayden,
Dear Sir,

I received the box containing saurian remains sent from St. Louis, and I immediately acknowledged its receipt, which as you were shortly afterwards in Washington, you did not get.

Among the remains were a parcel of fragments of dense rib-like bones with several fragmentary bodies of vertebrae. These I suspect to have belonged to a herbivarous cetacean of remarkable character. In what formation were they found?

The vertebra and other bones embedded in two large masses of hard matrix is a new and very peculiar genus of ichthyoid reptilia.

[20] A pioneer description of the prehistoric animals roaming the American continent.

I am sorry the matrix is so hard, as it might be impossible to remove it without breaking the fossils themselves. . . .

With respect, I am at your service,

Joseph Leidy

Philadelphia February 25, 1856

Where is the Judith River of which you speak? A branch of what larger stream is it?

L.

Leidy to Hayden

Philadelphia January 12, 1857

Dear Sir,

I have received yours of the 7th and 9th.

In regard to your collection after some conversation with several members of the Academy I am at liberty to offer you for it, the money intended towards an expedition, i.e. the sum of $1200, of which $600 will be paid cash when you arrive here and the remaining $600 will be subject to your order on receipt of the portion of the collection now in St. Louis. The Academy will also bear the freight of the collection.

In regard to publishing, I think you had better furnish me with some of your geological notes, and permit me to prepare a memoir for the Smithsonian immediately. One half of the copies given by the latter (100) to be at your service. The reason of the advice is as follows: Many of the animals, especially those of the Judith R. are characterised from a single tooth or often mere fragments, of which I have made drawings and descriptions. In your future explorations you may obtain more perfect material which then will render that at present possessed comparatively useless.

In regard to the figures for Lieut. Warren's[21] Report they will at least fill a quarto plate, which might be folded, or there must be two octavo plates.

In paying you the money for your collection instead of appropriating it to an expedition we think it is better for both parties, and your future paleontological collection if sent to the Academy, will most probably also be purchased, or at least we will do all we can to obtain for you a pecuniary equivalent for your valuable labors. . . .

Joseph Leidy

[21] Gouverneur Kemble Warren (1830–82), an Army engineer active in the West.

Leidy to Hayden

Philadelphia May 18, 1857

Dr. Hayden.

Dear Sir,

I have written to Shuman in relation to the specimens in my possession to be divided between the two Academies.

Enclosed are two letters, one of which was addressed to me though evidently intended for you. The Geolog. paper alluded to was sent and is now in my possession and if you wish it I will transmit it to you.

The British Government has formed an exploring party to determine a route to Vancouver's Island. They are to go west from the head waters of Lake Superior, and thus they may walk right into the affections of your Miss Judith. What is to be done.

With respect I am at your service.

Joseph Leidy

Leidy to Hayden

Philadelphia January 17th 1858

Dear Dr.

I have commenced examining the fossils last sent, and as you anticipated, I have been surprised. The fauna is so different from that of the Mauvaises Terres; the types of animals being more ancient than the present though less so than the former. Among the fossils were some fragments of jaws which I had referred to the camel tribe, and this result was confirmed by the portion of a lower jaw yesterday given to me by Dr. Hammond.[22]

The collection contains fragments of about twenty different animals, all I think different from those of the Nebraska Miocene; but some like those of Bijm Hills. I would ask if there is not a formation similar to the latter and that of Running Water at Bear Creek? I suspect so from some of the fossils.

If you have obtained anything further to assist me in my determination, I hope you will send them in at your early convenience through Adams Express.

This new pliocene collection only makes me thirst for more, as there are many fragments of carnivera and pachyderms quite evident but not sufficiently characteristic for description. I hope you

[22] William Alexander Hammond (1828–1900), then an Army surgeon; Surgeon General during the Civil War and afterwards a leading neurologist.

have it in your heart again to visit this locality as well as the one of Judith River. I have collected on your account $50. of which I paid to the printers as you directed 46.50. Of this Lieut. Warren recently sent $27.50 which I have put down to your credit.

With respect I am at your further service

Joseph Leidy

Leidy to Hayden

Philadelphia June 3, 1859

My dear Dr. Hayden,

I am exceedingly sorry not to have been able to send you a proof of your paper. The stupid printer has given me an immense deal of trouble with his delays and mistakes, and even now I have not yet been able to get a proof of your paper. Instead of giving me the latter he sent one of the first part of my paper, and when I informed him that you came first, he set up the proof of an Italian paper. I shall however carefully attend to sending a proof to Mr. Meek as well as read it myself. I ordered for you 100 extra copies, 200 for myself of the combined papers, with a common title page.

I hope you may have your usual good luck in rich paleontological discoveries, as also in everything else you undertake.

Hoping to hear from you, I remain your friend,

Joseph Leidy

P.S. Permit me to advise you again not to load yourself with huge turtles, fragments of the same, and fragments of the bones of the extremities of animals, as they are comparatively worthless. Of mammals confine your collections to skulls, teeth and fragments of the same, and only under peculiar circumstances preserve other bones. Of reptiles all vertebrae and articular fragments are of value, and of fishes the whole body.

Freights are enormous, and sometimes we are compelled to pay dollars for what are not worth as many cents.

L.

Leidy to Hayden

Philadelphia February 2, 1860

Dear Dr.

I have just received yours of Jan. 24. It is the third I have had the pleasure of receiving. In regard to the coloring of the map in our paper, I am sorry to say it was forgotten or overlooked by the publication committee. It was done in my absence from the city; at a time when I did not anticipate the issue of the Transactions. I regretted it, but found consolation in the fact that it was really of little consequence, as the boundaries of the formation are marked, and letters indicate their character.

In regard to my counterpart of whom you ask, I can only answer as perhaps you would, she is yet to be found. Miss P. is now actually married. Dr. Newberry[23] is now in Washington. He has recently sent to me for examination the remains of a huge reptile which he discovered in the Jurassic rocks of south Utah. They puzzle me, quite as much as they did Dr. N. which is what you would have suspected, for from what I have seen of the Dr. he is quite as able as I am to determine such matters. Dr. Evans is publishing the subject of his reports vigorously, but what may be the result I am unable to say. Dr. H. Engelmann recently sent me some remains collected in Capt. Simpson's[24] expedition, but nothing different from what you had already sent to me. Lieut. Warren passed through here on his way to Washington, and took with him my report on the Niobrara fossils. In my paper I refer to you for the geological information. I have not learned the result of his mission. . . .

Joseph Leidy

Hayden Papers, National Archives

The Coming of Natural Selection

Except for his voyage around the world in the *Beagle* as a young man, Charles Darwin stayed close to home, reading books, journals, and quietly gathering information by correspondence. Long before the publication of *The Origin of Species* in 1859 he had developed many of the ideas in that influential book. Darwin was not one to rush into print with every little fact or idea; nor was he the sort of scientist who glories in controversy. Shy, almost a recluse, Darwin believed in his insight into nature and marshaled data to back his beliefs. The opposition to evolution or "progressive development" was great and not entirely confined to theologians and to arguments drawn from the Bible. Arrayed against Darwin was an imposing body of scientists and of scientific thought. To overcome his

[23] John Strong Newberry (1822–92), later a professor at Columbia University.

[24] Engelmann was the geologist son of George Engelmann. Capt. James H. Simpson (1813–83) was another scientifically inclined Army engineer.

opponents, both scientific and theological, Darwin's strategy called for an overwhelming weight of evidence. Only the unexpected appearance of Alfred Wallace (1823–1913) with an independently conceived theory of natural selection jogged Darwin into publishing his book.

Of Darwin's American correspondents the two most important were Asa Gray and the geologist James Dwight Dana. They were leaders in their respective fields, not only nationally but internationally. As sources of information, they were valuable to Darwin, but their significance as correspondents went far beyond that role. Both Gray and Dana were orthodox, devout Christians. Gray soon became an adherent of Darwin; eventually Dana followed the example of his friend. Both men believed that Darwin should get a fair hearing in America on the scientific merits of his ideas and not be denied a fair chance on the grounds of irreligiousness. Dana, as the editor of *Silliman's Journal,* provided a forum for Gray but also for Louis Agassiz (1807–73), the principal American opponent of evolution. He himself was in very poor health and did not enter the fray. Gray could support natural selection in terms palatable to many believing Christians.

Agassiz' presence in America converted the controversy over evolution in the United States from what might have been a provincial sideshow to a major incident in one of the great intellectual controversies of the nineteenth century. Undoubtedly, many scientific worthies in America would have crossed pens with Gray. None of these could compete in popular esteem or professional renown with the Swiss scientist who had resided in America since 1846. Before crossing the ocean he had written a great work on fossil fish and had pioneered in the development of the theory of the glacial periods or "ice ages" when much of the temperate zone was covered by a sheet of ice. Agassiz was the friend, protégé, and associate of many of the great names in European science. As a sincere believer in the permanence of species, he could not, at first, even bring himself to consider seriously Darwin's facts or even the kind of arguments used by Darwin.

Like Gray, Agassiz was at Harvard. At first both were fairly friendly. By 1859 many differences existed between the two scientists. The evolution controversy only widened the rift. Gray was notably excluded from the dinners of the Lazzaroni. At Harvard the two men and their allies clashed over educational policy. To personal animosity and institutional and group rivalries were now added ideological bitterness.

Redfield to Louis Agassiz

New York 23rd December 1846

My dear Sir

I received yesterday your enclosure containing the Monograph of Old Red Fishes, consisting of 3 number of text and three of plates, together with a number on Echinodermes, all apparently in good order. I do not know the price, but if you will inform me I will remit the amount to you at Boston, or pay it to you here on your arrival, as may best suit your convenience. I had before seen only the figures in Mr. Millers book,[25] and am truly gratified with this wonderful display of creative power in the fishes in this ancient formation. It reminds me, also, that I have a few vestiges of [one word unclear] etc. which I do not remember to have shown you. Your beautiful drawings and restorations also make me anxious to see the development of our own triassic fishes, according to the plan which you were pleased to adopt when with us in New York. I suppose that about the 1st of Feb. will be as convenient a time for us here as any other. I shall probably go to Albany before that time, and if I accomplish my business there in suitable time, it will delight me to meet you at Boston and go with you to Amherst and Greenfield, so as to take charge *ourselves* of such specimens of ichthyolites as may be borrowed from the cabinets of President Hitchcock and Mr. Marsh. I know that President H. desires much to see you, and you will be much interested with the zoological vestiges of the Triassic rocks which they and Dr. Dean have collected.[26] You will doubtless have understood that it is the fixed purpose and desire of my son and myself that you should consider the little which we have done or attempted to do as being placed wholly at your disposal, as well as all other facilities which our little domestic establishments or our personal exertions are capable of affording for the execution of your labors; and in these matters we beg you to act with the greatest freedom and confidence.

I learned that arrangements are progressed for inviting you to deliver a course of lectures in this city during the present season, which I hope may be brought about in a manner consistent with your convenience. Barren as this particular region must appear, in the cultivation of Natural History etc. science, we have doubtless a multitude of minds which need only proper excitement and direction to bring them forward as successful laborers in this glorious field. Even in our Lyceum of Natural History we have some 8 or 10 young men whom your countenance and instructions might serve to place fairly on their feet, and might enable them hereafter to perform good service in the cause of science.

[25] Hugh Miller (1802–56), English geologist. This reference is probably to *Old Red Sandstone,* 1840.

[26] Edward Hitchcock (1783–1864), president of Amherst, geological theologian; Dexter Marsh (1806–53); and James Deane (1801–58). All three were involved in the discovery of fossil footprints.

Trusting that you will take good care not to be overworked in a country where so much is required to be done, I am dear Sir faithfully and truly yours

Wm. C. Redfield

[P.S. omitted]

Redfield to Agassiz

New York 26th January 1847

Dear Sir,

By mistake I have brought away your ticket for the Astronomical lectures, which you will find enclosed.

I learned last evening at the meeting of the Lyceum of Natural History that the gentlemen who are canvasing for a class or audience for the Course of Lectures which it is hoped you will favor us with, have engaged about 250 tickets, predicated on a course of 12 lectures. And that they have no doubt of engaging as many as the lecture room proposed will conveniently accomodate. Our President, Maj. [John] Delafield, who is one of the Trustees of College of Physicians and Surgeons also informed us that the board of Trustees had voted the free use of the College lecture room for this object. In view of these statements our Lyceum framed a Resolution requesting the President to invite you to deliver the Course, and I suppose that a like invitation will also be given by the Faculty of the College of Physicians and Surgeons. These invitations I think you will receive soon, and we earnestly hope that you will find it convenient to accept them.

The class or audience contemplated will probably comprise those persons in our city who are most fond of the natural Sciences, but will necessarily be somewhat promiscuous and will probably include some ladies, unless this should be deemed objectionable. I apprehend that the pecuniary conditions will have no particular influence with you, but it is proper to state that the plan pursued thus far is predicated on the purchase of the tickets at 3 dollars each. I do not know but that the expense of light and attendents will have to be deducted from the avails. The matter, I believe, is chiefly in the hands of gentlemen acting voluntarily, among whom are Dr. LeConte,[27] Dr. Adams,[28] and others of our Lyceum. As regards the character of the desired course the gentlemen do not seem to be acquainted with your views and preferences, but I

[27] Probably John Lawrence LeConte (1825–83), the entomologist.

[28] Dr. John Glover Adams (1807–44).

presume they are expecting something of the kind and title of your Lowell lectures at Boston.

I hope that the desired object can be attained, in some manner convenient to yourself. And if you can instruct me in any way that will enable me to promote your views, in any respect, it will afford me great satisfaction.

With my respects to Count Portales,[29] I am Dear Sir truly yours

Wm. C. Redfield

P.S. I brought from New Haven yesterday the specimens of fossil fishes which you saw at the rooms of the Yale Natural History Society which include the Catapteries from which my son's drawing was taken.

W.C.R.

Redfield Papers, Yale University

Charles Darwin to Asa Gray

Down Farnborough Kent April 17th 1855

My dear Sir

I hope that you will remember that I had the pleasure of being introduced to you at Kew. I want to beg a great favor of you, for which I well know I can offer no apology. But the favor will not cause you much trouble and will greatly oblige me. As I am no Botanist, it will seem so absurd to you my asking botanical questions, that I may premise that I have for several years been collecting facts on "Variation," and when I find that any general remark seems to hold good amongst animals, I try to test it in Plants. I have the greatest curiosity about the alpine Flora of the U. S., and I have copied out of your Manual the enclosed list; now I want to know whether you will be so very kind as to append from memory (I have not for one instance the presumption to wish you to look to authorities) the other habitats or range of these plants: appending "Indig." for such as are confined to the mountains of the U. S. "Arctic Am." to such as are also found in Arctic America. "Arctic Eup." to those also found in Arctic Europe, and "Alps" to those found on any *mountains* of Europe "and Arct. Asia" I have compared to list with the plants of Briton, but I am afraid of trusting to myself, from ignorance of geography.

I see that there are 22 species common to the White Mt. and

[29] Louis François de Pourtales (1823–80), a marine zoologist who came to America with Agassiz. For many years Pourtales worked at the Coast Survey.

the Mt. of New York, will you tell me about how wide a space of low land, on which these alpine plants cannot grow, separates these mountains: I can hardly judge from the height not being marked on the publication of the mountains of Vermont.

I return to ask for one more piece of information, viz. whether you have anywhere published a list of the phenerogamic species common to Europe, as has been done with the shells and Birds, so that a non-Botanist may judge a little on the relationship of two places. Such a list would be of extreme interest for me in several points of view and I should think for others. I suppose there would not be more than a few hundred out of the 2004 species in your Manual. Should you think it very presumptuous in me to suggest to you to publish (if not already done) such a list in some Journal? I would do it for myself but I should assuredly fall into many blunders. I can assure you, that I perceive how presumptuous it is in me, not a Botanist, to make even the most trifling suggestion to such a Botanist as yourself; but from what I saw and have heard of you from our Dear and Kind friend Hooker,[30] I hope and that you will forgive me and believe me, with much respect,

Dear Sir
Yours' very faithfully
Charles Darwin

Gray Papers, Harvard University

Gray to Darwin[31]

Harvard University, Cambridge June 30, 1855

Your long letter of the 8th instant is full of interest to me, and I shall follow out your hints as far as I can. I rejoice in furnishing facts to others to work up in their bearing on general questions, and feel it the more my duty to do so inasmuch as from preoccupation of mind and time and want of experience I am unable to contribute direct original investigations of the sort to the advancement of science.

Your request at the close of your letter, which you have such needless hesitation in making, is just the sort of one which it is easy for me to reply to, as it lies directly in my way. It would

[30] Sir Joseph Dalton Hooker (1817–1911), the English botanist.
[31] Their correspondence started, Darwin expanded his dialogue with Gray from geographic distribution to the question of the relationship of the number of close species to the size of their populations.

probably pass out of my mind, however, at the time you propose, so I will attend to it at once, to fill up the intervals of time left me while attending to one or two pupils. So I take some unbound sheets of a copy of the *Manual,* and mark off the "close species" by connecting them with a bracket.

Those thus connected, some of them, I should in revision unite under one, many more Dr. Hooker would unite, and for the rest it would not be extraordinary if, in any case, the discovery of intermediate forms compelled their union.

As I have noted on the blank page of the sheets I send you (through Sir William Hooker), I suppose that if we extended the area, say to that of our flora of North America, we should find that the proportion of "close species" to the whole flora increased considerably. But here I speak at a venture. Some day I will test it for a few families.

If you take for comparison with what I send you, the *British Flora,* or Koch's *Flora Germanica,* or Godron's *Flora of France,* and mark the "close species" on the same principle, you will doubtless find a much greater number. Of course you will not infer from this that the two floras differ in this respect; since the difference is probably owing to the facts that (1) there have not been so many observers here bent upon detecting differences; and (2) our species, thanks mostly to Dr. Torrey and myself, have been more thoroughly castigated. What stands for one species in the *Manual* would figure in almost any European flora as two, three, or more, in a very considerable number of cases.

In boldly reducing nominal species J. Hooker is doing a good work; but his vocation like that of any other reformer exposes him to temptations and dangers.

Because you have shown that a and b are so connected by intermediate forms that we cannot do otherwise than regard them as variations of one species, we may not conclude that c and d, differing much in the same way and to the same degree, are of one species, before an equal amount of evidence is actually obtained. That is, when two sets of individuals exhibit any grave differences, the burden of proof of their common origin lies within the person who takes that view; and each case must be decided on its own evidence, and not on analogy, if our conclusions in this way are to be of real value. Of course we must often jump at conclusions from imperfect evidence. I should like to write an essay on species some day; but before I should have time to do it, in my plodding way, I hope you or Hooker will do it, and much better by far. I am most glad to be in conference with Hooker and yourself on these matters, and I think we may, or rather you may, in a few years settle the question as to whether Agassiz's or Hooker's views are correct; they are certainly widely different.

Apropos to this, many thanks for the paper containing your experiments on seeds exposed to sea water. Why has nobody thought of trying the experiment before, instead of taking it for granted that salt water kills seeds? I shall have it nearly all reprinted in Silliman's Journal as a nut for Agassiz to crack.

More Letters of Charles Darwin, vol. 1, pp. 421-22

Darwin to Gray

Down Bromley Kent July 20 [1856]

My dear Dr. Gray:

What you say about extinction, in regard to such genre and local disjunction, being hypothetical seems very just. Something direct however, can be advanced on this head from fossil shells; but hypothetical such notions must remain. It is not a little egotistical, but I should like to tell you, (and I do not *think* I have) how I view my work. Nineteen years (!) ago it occurred to me that whilst attention employed on Natural History I might perhaps do good if I noted any sort of facts bearing on the question of the origin of species; and this I have since been doing. Either species have been independently created, or they have descended from other species, like varieties from one species. I think it can be shown to be probable that man gets his most distinct varieties by preserving such as arise best worth keeping and destroying the others, but I should fill a quire if I were to go on. To be brief I *assume* that species arise like our domestic varieties with *much* extinction; and then test this hypothesis by comparison with as many general and pretty well established propositions as I can find made out, in geographic distribution, geological history, affinities etc. etc. etc. And it seems to me, that *supposing* that such hypotheses were to explain such general propositions, we ought in accordance with common way of following all sciences, to admit it, till some better hypothesis be found out. For to my mind to say the species were created so and so is no scientific explanation but a prescient[ific] way of saying it is so and so. But it is not sensible trying to show how I try to proceed in compass of a note. But as an honest man I must tell you that I have come to the relentless conclusion that there are no such things as independently created species, the species are only strongly defined varieties. I know that this will make you despise me. I do not much under-rate the many *huge* difficulties on this view, but yet it seems to me to explain too much, otherwise inexplicable, to be false. Just to allude to one point in your last note, viz about species of the same genus *generally* having a common or continuous area: if they are actual lineal descendents of one species, this of course would be the case; and the sadly too many exceptions (for me) have to be explained by climactic and geological changes. A fortiori on this view (but on exactly same grounds) all the individuals of the same species should have a continuous distribution. On this latter kind of subject I have put a chapter together and Hooker kindly read it over: I thought the exception[s] and difficulties were so great that on the whole the balance weighed against my notions, but I was much pleased to find that it seemed to have considerable weight with Hooker, who said he had never been so much staggered about the permanence of species. I must say one word more in justification (for I feel sure that your tendency will be to dispare over my contents) that all my notions about *how* species change are derived from long continued study of the works of (and concern with) agriculturalists and horticulturalists; and I believe I see my way pretty clearly on the means used by nature to change the species and *adapt* them to the conditions and exquisitely beautiful contingencies to which every living being is exposed.

I thank you much for what you say about variability and crossing of the grasses: I have been often astounded at what Botanists say on fertilization in the bud: I have seen *Cincifera* mentioned as instances, which every gardener knows how difficult it is to protect from cuping! What you say on Popilionaceous flowers is very true; and I have no facts to show the varieties are cuped; but yet (and the same remark is applicable in a beautiful way to Frumaria and Dielytia as I noticed many years ago) I must believe that the flowers are constructed partly in direct relation to insects' visits; and how insects can avoid bringing pollen from other individuals I cannot understand. It is really pretty to watch the action of a *Humble*-Bee on the scarlet Kidney Bean, and in this genus (and in Lathrus Grand. flowers) the honey is so placed that the Bee invariably alights on the side of the flower towards which the pistol is pointed (bringing out with it pollen) and by the depression of the wing-petal is forced against the Bees' side all dusted with pollen. In the Broom the pistol is rubbed on centre of back of Bee. I suspect there is something to be made out about the Leguminosae which will bring the case within *our* theory: though I have failed to do so. For theory will explain why in vegetable and animal Kingdoms the act of fertilization even in hermaphrodites usually takes place sub-jove, though thus exposed to the *great* injury from damp and rain. In animals in which the *semen* cannot, like pollen be *occasionally* carried by insects or wind: there is *no case* of *Land*-animals being hermaphrodite without the concourse of two individuals. But my letter has been horribly egotistical: but your letters always so greatly interest me; and what is more they have in simple truth, been of the *utmost* value to me.

Yours most sincerely and gratefully
C. Darwin

N.B if you will look at bed of Scarlet Kidney Bean you will find that the wing-petals on the *Left*-side alone are all scratched by the tassi of the Bees.

Gray Papers, Harvard University

Darwin to Gray

Down Bromley Kent Sept. 5 [1857]

My dear Gray:
I forget the exact words which I used in my former letter, but I daresay I said that I thought you would utterly despise me, when I told you what views I had arrived at, which I did because I thought I was bound as an honest man to do so. I should have been a strange mortal, seeing how much I owe to your quite extraordinary kindness, if in saying this I had meant to attribute the least bad feeling to you. Permit me to tell you, that before I had even corresponded with you Hooker had shown me several of your letters (not of a private nature) and these gave me the warmest feeling of respect to you; and I should indeed be ungrateful if your letters to me and all I have heard of you, had not strongly enhanced this feeling. But I did not feel in the least sure that when you knew whither I was tending, that you might not think me so wild and foolish in my views (God knows arrived at slowly enough, and I hope conscientiously) that you would think me worth no more notice or assistance. To give one example, the last time I saw my dear

old friend Falconer, he attacked me most vigorously, but quite kindly, and told me "you will do more harm than any ten naturalists will do good." "I can see that you have already corrupted and half-spoiled Hooker" (!!) Now when I see such strong feeling in my oldest friends, you need not wonder that I always expect my views to be received with contempt.

But enough and too much of this. I thank you most truly for the kind spirit of your last letter. I agree to every word in it; and think I go as far as almost anyone in seeing the grave difficulties against my doctrine. With respect to the extent to which I go, all arguments fade *tepidly* away the greater the scope of forms considered. But in animals, embryology leads us to an enormous and frightful range. The facts which kept me longest scientifically orthodox are those of adaptation, the pollen masses in Asclepias, the [mistletoe with its pollen carried by insects], the woodpecker with its feet and tail beek and tongue to chink trees and secure insects. To talk of climate or Lamarckian habit producing such adaptations to other organic beings is futile. This difficulty, I believe I have surmounted. As you seem interested in subject, and as it is an *immense* advantage to me to write to you and to hear *ever so briefly,* what you think, I will enclose (*copied* so as to save you trouble in reading) the briefest abstract of my notions as to *means* by which nature makes her species.[32] Why I think the species have really changed depends on general facts in the affinities, embryology, rudimentary organs, geological history and geographical distribution of organic beings. In respect to my abstract you must take immensely on trust; each paragraph occupying one or two chapters in my Book. You will, perhaps, think it paltry in me, when I ask you not to mention my doctrine; the reason is, if any one, like the author of the Vestiges,[33] were to hear of them, he might easily work them in, and then I should have to quote from a work perhaps despised by naturalists and this would greatly injure any chances of my views being received by those alone whose opinions I value. I have been lately at work on a point which interests me *much;* namely dividing the species of several Floras into two as nearly as equal cohorts as possible, one with all those forming larger genera, and the other with the smaller genera. Thus in your United States Flora,

[32] Not printed here.

[33] *The Vestiges of Creation,* published anonymously in 1844 by Robert Chambers, created a sensation by its espousal of evolution, but its crudities were roundly scored by many naturalists.

I make (with omissions of naturalized and a few protean genera and Carex for its annual size) 1005 sp[ecies] in genera of 5 and upwards, and 917 in genera with 4 and downwards; and the larger genera have 88/1000 varieties and the smaller genera only 50/1000. This rule *seems* to be general and Hooker is going to work out some others on same plan. But to my disgust your *var.* markedly big-typic are only in proportion 48/1000 to 46/1000. Several things have made me confidently believe that "close" species occurred most frequently in the larger genera; and you may remember that you made me the enclosed list. Now to my utter disgust, I find that the case is somewhat the reverse of what I had so confidently expected. The close species hugging the smaller genera. Hence I have enclosed the list and beg you kindly to pass your eye over it, and see whether, not understanding my notice, you could have attended more to the smaller than to the larger genera: but I can see that this is not probable. And do not think that I want you to "cook" the results for me. Are the close species *less generally* geographical representation species: this might make some difference? Lately I examined *buds* of Kidney Bean with pollen shed, but I was led to believe that the pollen could *hardly* get on stigma by wind or otherwise, except by Bees visiting and moving the wing pellets: Hence I included a small bunch of flowers in my two bottles, in every way treated the same: the flowers in one I daily just momentarily moved as if by a Bee; these set 3 fine pods, the other not *one*. Of course the little experiment must be tried again, and this year in England it is too late, as the flowers seem but seldom to set. If Bees are necessary to this flower's *self*-fertilization, Bees *must* almost cross them, as their dusted right-side of head and right legs constantly touch the stigma. I have, also, lately been reobserving daily Lobelia fulgens, this in my garden is never visited by insects and never sets seeds, *without pollen be put on* stigma. (whereas to the blue Lobelia is visited by Bees and does set seed); I mention this because these are such beautiful contrivances to prevent the stigma ever getting its own pollen; which seems only explicable on the doctrine of the advantage of cupes.

I forget whether I ever said I had received safely Mr. Watsons papers and your "Lesson in Botany", for which my many thanks and which I am now reading. But I have even had the last part of your paper on Naturalized Plants. If you have a spare copy (which is not likely) I should be very glad of it: otherwise I will borrow Hooker's. I ought to feel ashamed at the thought of the latter knowing how busy you are. My dear Dr. Gray believe me with much sincerity. Yours truly

C. Darwin

I will try if I can anyhow get seed of the Adlumia curhosa and observe it next summer. Perhaps they have it Kew.

Gray Papers, Harvard University

Gray to James Dwight Dana

Cambridge 7th Nov. 1857

My Dear Dana

If you have plenty, please send me 2 more copies of your *Thoughts on Species*[34]

I first read it carefully, a week ago, and I meant to write to you at once how I like it, and a few remarks, but something prevented at the time, and I have been very busy and preoccuppied ever since.

For the reason I like the general doctrine, and wish to see it established, so much the more I am bound to try all the steps of the reasoning, and all the facts it rests on, impartially, and even to suggest all the adverse criticism I can think of. When I read the pamphlet I jotted down in the margin some notes of what struck me at the time. I will glance at them again, and see if, on reflection they appear likely to be of the least use to you, and if so will send them, taking it for granted that you rather like to be criticized, as I am sure I do, when the object is the surer establishment of truth.

In your idea of species as specific amount and/or kind of concentered force, you fall back upon the broadest and most fundamental view, and develop it, it seems to me, with great ability and cogency.

Taking the *cue* of species, if I may so say, from the *inorganic* you develop the subject to great advantage for your view, and all

[34] Published in 1857 in *Silliman's Journal.* Dana argued for the permanence of species. His argument was largely based on analogy with inorganic nature. In his classic *System of Mineralogy,* the first edition of which appeared in 1837, Dana first arranged the minerals as in a natural history classification with classes, orders, and species, only dropping the scheme in his 3rd ed. of 1850. Dana, the orthodox scientist, was invoking a kind of materialism to support antievolutionary views. Gray, who had already read Darwin's thesis transmitted in the letter of Sept. 5, 1857, raised a kind of vitalistic argument in opposition but did not openly espouse evolution.

you say must have great weight, in "reasoning from the general".

But in reasoning from *inorganic species* to organic species, and in making it tell *where* you want it, and *for what* you want it to tell, you must be sure that you are using the word *species* in the same sense in the two—that the one is really an equivalent of the other. That is what I am not yet convinced of. And so to me the argument comes only with the force of an *analogy*, whereas I suppose you want it to come as demonstration. Very likely you could convince me that there is no fallacy in reasoning from the one to the other, to the extent you do. But all my experience makes me cautious and slow about building too much upon analogies; and until I see further and clearer, I must continue to think that there is an essential difference between *kinds of animals or plants* and *kinds of matter*. How far we may safely reason from the one to the other is the question. If we may do so even as far as you do, might not Agassiz (at least plausibly) say, that as the *species Iron* was created in a vast number of individuals over the whole earth, so the presumption is that any given species of plants or animals was originated in as many individuals as there are now, and over as wide an area, the human species under as great diversities as it now has (barring historical intermixtures), and so reducing the question between you to insignificance? Because then the question whether men are of one or of several species would no longer be a question of fact, or of much in science.

You can answer him from *another starting point*, no doubt: but he may still insist that it is a legitimate carrying out of your own principle.

P. 307: line 11 from bottom, after "group" at the end of the line add, nor in the series.

And in one just sense, I agree that "the true notion of the species is not in the resulting group," nor in the genetic series. But objectively the organic species is realized or expressed in the series (as well as in the cyclical changes). And even subjectively, your initial being is *individual:* and the idea of *species* wholly grows out of the difference between this initial individual and other initial individuals, i.e. out of *original diversity of kind*, which we suppose to have been *ab initio*.

P. 308. line 9 from bottom. For "necessarily" read probably. "Necessarily" begs a question, and closes the argument. It is inferring from the inorganic to the organic with the force of a *Q.E.D.* Your "necessarily" is merely *rhetorical*, I suppose. There are first several more or less probable suppositions to be eliminated, and analogies to be considered, and some spirits of your own raising may possibly need to be more thoroughly laid. As the species Sulphate of Lime is composed of two other species, and these each of two others, may not the analogue of this in the organic world (where *reproduction* as you neatly say takes the place of *combination*) be, that species procreate species, and so increase the kinds? And in the capacity of every species that we have experimentally tested to develop races, and in the tendency of races to fixity under favorable circumstances may we not have some evidence of this? The tendency of my mind is opposed to this sort of view: but you may be sure that before long there must be one more resurrection of *the development* theory in a new form, obviating many of the arguments against it, and presenting a more respectable and more formidable appearance than it ever has before.

P. 309, line 1. What if they blend definitely? so to say, or rather, to make a supposition more accordant with the baring of the facts —what if many or some species vary until (however distinct *ab initio*) they blend, as to all external appearances with their congeners?

[P. 309,] line 8, 9, 10. As to genera, etc. it seems to be very much so, a *superficial glaring over the surface,* still giving us the inevitable idea of definite elements in the structure, but leaving much of the mode and extent of the combinations dim, and to be sought after, *haply* we may find them.

[P. 309,] line 20 from bottom. at end, for "the" read *either.*

I have no notion of hybridity being an active cause of variation in matter, but it is quite possibly active to some extent, in a way which is not provided for, as I see, in either of your seven categories. Let me add it, as 4½. mules that are indefinitely fertile in their descendents fertilized in their first generation by one of their parents. This is generally conceded to happen in cultivated plants, and may happen in wild. They say the progeny reverts to the type of that parent: but it carries with it a dash of foreign blood, and so must be admitted as a cause of variation, so far as it goes tending towards blending related species. It affords the most probable explanation of remarkable intermediate forms between certain species of oak, for instance.

P. 310. line 22, 23, hardly enough "to cut the hypothesis short"; for there are many species more nearly related than the *Horse* and the *ass.*

And as to the end of that paragraph, it is not reasonable, or rather not conclusive, to assume that the general spirit of Nature's system, which rests only on general inferences, forbids a supposition in a particular case which you say "it is fair to make." This, however is petty criticism, of the form or language not of the substantial subject matter.

The paragraph at the foot of p. 310 is *capitally put, it tells through and through!* And I see no way of avoiding its force.

Somewhere here or before, I should like to bring in my idea of cross breeding between individuals of a species (*in plants*), which in so many cases remarkably provided for in Vegetable Kingdom, *as a natural agency for repressing development of races,* and which would obliterate any existing ones where it had full play. I have referred to it in the Journal once or twice.

P. 311, line 11 from bottom "Diverge," yes, So widely *in genera,* that you cannot safely argue far from the one to the other.

P. 314, line 3 et seq. Variations of condition are here

compared — variations
with — or — of *character.*
confounded ? — differences

I wanted to say something on the last two pages, but as I have nothing in particular to except to, and much to approve. And as it is late bed-time. I spare you further comments.

I set out to find *flaws,* as likely to be more suggestive and there fore far more useful to you than any amount of praise, with which I can fill page after page.

Ever Yours Sincerely,
A. Gray

Dana Papers, Yale University

Darwin to Agassiz

Down Bromley Kent Nov. 11th [1859]

My dear Sir:

I have ventured to send you a copy of my Book (as yet only an abstract)[35] on the origin of species. As the conclusion at which I have arrived on several points differ so widely from yours, I have thought

[35] Whether out of modesty or caution, Darwin kept pretending that his solid book and his other publications were only abstracts for a full, definitive treatise.

(should you at any time read my volume) that you might think that I had sent it to you out of a spirit of defiance or bravado; but I assure you that I act under a wholly different frame of mind. I hope that you will at least give me credit, however erroneous you may think my conclusion, for having earnestly endeavoured to arrive at the truth.

With sincere respect, I beg leave to remain

Yours very faithfully
Charles Darwin

Louis Agassiz Papers, Harvard University

Gray to Francis Boott[36]

Cambridge 16th January 1860

My Dear Boott

Darwin's book reached me about Christmas, and I had read it carefully by New Year or soon after.

Agassiz cannot abide it (of course) and so has publicly denounced it as atheism etc. etc.

I am bound to stick up for its philosophy, and I am struck with the great ability of the book and charmed with its fairness, I also wanted to stop Agassiz's mouth with his own words, and to show up his loose way of putting things. He is a sort of demogogue, and always talks to the rabble.[37]

So I have written a long article in Silliman on Darwin, to be out 6 weeks hence, when I will send you a copy.

It has taken a solid fortnight from the time I took pen in hand . . .

Ever yours affectionately,
Asa Gray

Darwin-Lyell Papers, American Philosophical Society

Darwin to Gray

Down Bromley Kent February 18th [1860]

My dear Gray:

[36] Francis Boott (1792–1863) was an American-born physician who lived in England. He was a skillful botanist.

[37] i.e., Agassiz was taking the matter to the people. The Darwinists and some scientists hostile to Darwin preferred to have the controversy decided within the scientific community on scientific merits. Great issues, however, defy confinement to pigeonholes.

I received about a week ago two sheets of your Review; read them, and sent them to Hooker; they are now returned and re-read with care, and tomorrow I send them to Lyell.

Your Review seems to me *admirable:* by far the best which I have read. I thank you from my heart both for myself, but far more for subject-sake. How curious your contrast between the views of Agassiz and such as mine is very curious and instructive. By the way if Agassiz writes anything on subject, I hope you will tell me. I was charmed with your metaphor of the streamlet never running against the force of gravitation. Your distinction between an hypothesis and theory seemed to me very ingenious; but I do not think it is ever followed. Everyone now speaks of the undulatory *theory* of light; yet the ether is itself hypothetical and the undulations are inferred only for explaining the phenomena of light. Even in the *theory* of gravitation, is the attractive power in any way known, except by explaining the fall of the apple and the movements of the Planets?

It seems to me that an hypothesis is developed into a theory solely by explaining an ample lot of facts. Again and again I thank you for your generous aid in discussing a view, about which you very properly hold yourself unbiased.

My dear Gray
Yours most sincerely
C. Darwin[38]

Gray Papers, Harvard University

Excerpt from Benjamin Peirce's 1860 Journal of a European Trip[39]

. . . There has been a vast [gap in mss.] of discussion in the other section about Darwin's book, which is occupying all the attention of England. I heard a very sharp pass between Owen and Huxley, and a long and very earnest one between the Bishop of Oxford and Huxley,[40] at which I should think that there must have been a thousand persons present. The Bishop is one of the most eloquent men I ever heard, he's known here as Soapy Sam, and the slippery character of the divine was apparent in all his argument. His power of language was wonderful and the revulsion? with which

[38] Postscript omitted.
[39] Entry for June 30, 1860.
[40] Samuel Wilberforce, Bishop of Oxford (1805–73) and Sir Richard Owen (1804–92) a zoologist, represented orthodox religion and science, respectively. Thomas Henry Huxley (1825–95) became the great public champion of Darwin

he seized upon the weak points of his opponents views and exposed them to the torture was a model of logical display. . . .[41]

Benjamin Peirce Papers, Harvard Archives

Darwin to Dana

Down Bromley Kent July 30th [1860]

My dear Sir:

I received several weeks ago your note telling me that you could not visit England, which I sincerely regretted, as I should most heartily like to have made your personal acquaintance. You gave me an informed, but not very good, account of your health. I should at latter times be gratified for a letter to tell me how you are. We have had a miserable summer owing to a terribly long and serious illness of my eldest girl, who improves slightly but is still in a precarious condition. I have been able to do nothing in science of late. My kind friend Asa Gray often writes to me and tells me of the warm discussions on origin of species in the United States. Whenever you are strong enough to read it, I know you will be dead against me, but I know myself well that your opposition will be liberal and philosophical. And this is a good deal more than I can say of all my opponents in this country. I have not yet seen Agassiz's attack; but I hope to find it at home, when I return in few days, for have been for several weeks away from home on my daughter's account. Prof. Silliman sent me an extremely kind message by Asa Gray that your Journal would be open to a reply by me: I cannot decide till I see it, but in preparation I have decided to avoid answering anything, as it consumes much time, often temper, and I have said my say in the Origins. No one person understands my views and has defended them so well as A. Gray; though he does not by any means go all the way with me. There was much discussion on subject at British Association at Oxford; and I had my defenders and my side seems (for I was not there) almost to have got the best of the battle. James Comerford and my neighbor J. Lubbock goes on working at such spare time as

[41] Peirce, a friend of Agassiz and inclined to idealism in philosophy, was clearly not impressed by the evolutionist's arguments. An American, John William Draper the chemist, provided the occasion for the now famous debate between Wilberforce and Huxley. Draper, an ardent believer in the idea of progress and in a kind of religion of science, outraged the orthodox with his now forgotten talk, "On the Intellectual Development of Europe, considered with Reference to the Views of Mr. Darwin and Others that the Progression of Organisms is Determined by Law."

he has. This is an egotistical note; but I have not seen a Naturalist for months. Most sincerely and dearly do I hope that this note may find you almost recovered. Pray believe me

yours very truly,
C. Darwin

Dana Papers, Yale University

THE FOUNDING OF THE NATIONAL ACADEMY OF SCIENCES

In the middle of the Civil War Congress passed a bill establishing a National Academy of Sciences, which was duly signed by President Lincoln. Neither Congress nor President paid much attention to the act of March 3, 1863, whose sponsor, Senator Henry Wilson of Massachusetts, adroitly took advantage of the usual near chaos of the closing hours of a legislative session to push the bill through. The legislation was very simple. Section one named fifty Americans as the incorporators and founding members; section two limited the membership to fifty and authorized the Academy to elect members, divide into classes, and make all necessary rules; section three called for an annual meeting and authorized the Academy, upon the request of the Government, "to investigate, examine, experiment, and report upon any question of science and art." The Academy would be reimbursed for the actual expenses of such work, but its members would not receive any compensation for its services to the government.

Behind this seemingly innocent legislation was Alexander Dallas Bache and his devoted band, the Lazzaroni. Since at least 1851 they had worked for an organization of the leading scientific savants—a select group in marked contrast to the American Association for the Advancement of Science, which was open to all friends of science, whatever their actual attainments in research. As will appear from the letters that follow, the Lazzaroni were concerned with standards of professional competence in order to squeeze out and to put in their proper place those they considered inadequate as scientists. The Lazzaroni also sought a relationship with the government that would provide funds for science. Their beloved chief, Alexander Dallas Bache, naturally thought of federal funds since he was the best example of how a scientist might use such funds. All of the group probably were influenced by the precedent of European academies and learned societies with their support from royalty and the aristocracy. Finally, the Civil War engendered patriotic emotions, and the scientists were anxious to serve the Union.

When the question of aiding the Union came up in discussions among the Lazzaroni in Washington in the winter of 1862–63, Joseph Henry objected to the proposal of an Academy on the grounds, among other reasons, of political inexpediency and that it would arouse jealousies and antagonisms in the scientific community. Henry, Bache, and Charles Henry Davis (Chief of the Navy's Bureau of Navigation, the Department's main scientific organization) decided to promote the formation of a committee to advise the government on inventions and other scientific proposals. The result was the establishment of the Permanent Commission in the Navy Department on February 11, 1863, which functioned throughout the Civil War.

Unknown to Joseph Henry, Bache, Davis, the astronomer Benjamin Apthorp Gould, Benjamin Peirce, and Louis Agassiz continued to maneuver for the Academy. Senator Wilson was brought into the picture by Agassiz. The passage of the legislation was a surprise to almost all of the incorporators outside the small inner group. The original intention was for a small number of founders who would fill the remaining places by election. To expedite passage of the bill, however, all the places were filled at once. The general composition of the Academy reflected the views of the Lazzaroni; glaring omissions were widely and probably correctly ascribed to the personal animosities of the inner group. Many scientists, even among the chosen fifty, were doubtful or hostile to the new body. Until vacancies appeared, no new members could be chosen in order to assuage hostile opinion.

Of the fifty incorporators, thirty-two were in the physical sciences (mainly geophysics) and eighteen were in natural history. It is doubtful whether this proportion was a true measure of achievement between the two divisions; nor did it accurately reflect the proportion of scientific work performed in the United States. The natural historians were obviously being slighted. Twenty-two incorporators were in some way connected with the federal government. While the federal government certainly loomed large in the scientific scene, this high percentage reflected Bache's orientation.

At least three notable scientists were omitted. George P. Bond, the Director of the Harvard Observatory, had incurred the emnity of Pierce, Gould, and Bache. For his research Bond clearly merited inclusion ahead of others on the list; as he was at that time fatally ill with tuberculosis, many were outraged at this slight. Spencer F. Baird, the Assistant Secretary of the Smithsonian, was widely respected by the naturalists but not in the good graces of the great Agassiz. The chemist John William Draper had a world-wide reputation for achievements beyond those of all but a few of the incorporators. He was left out probably either because he simply did not fit into any of the conventional groupings of American scientists of that day, or because his views were such that they would have brought upon the Academy the hostility of the orthodox.

Everything went well for the Lazzaroni at first. The first meeting in 1863 elected Bache President of the Academy. Joseph Henry agreed to remain quiet in deference to his old friend. The principal opposition to him was displayed by William Barton Rogers (1804–82), the geologist who founded the Massachusetts Institute of Technology. But many others were also unhappy. By the 1864 meeting in New Haven the situation had changed drastically. Bache, the brain and muscle of the Lazzaroni, was incapacitated by a stroke and remained an invalid until his death in 1866. Asa Gray, who was contending with the Lazzaroni in Cambridge and particularly with Agassiz on evolution, was openly hostile to their ideas. He, Henry, and Dana joined forces to effect the election of Baird over the bitter opposition of Agassiz. Baird's membership represented both a repudiation of the personal animosities and also of the standards that the Lazzaroni were trying to impose.

The Academy might have died a lingering death afterward but for the influence of Bache, effective even from the grave. Bache willed his estate to the Academy, and Henry, out of deference for his friend's memory, took the Academy under his wing. He expanded the membership and transformed the organization into a learned society like the local groups already in existence. From the Bache estate and other funds it received the Academy could modestly support research and award medals and prizes. In time membership in the group became a recognized honor.

The Lazzaroni conception died hard. As late as 1892 Gould offered to give a fund to the Academy provided the membership was reduced from 100 to 50 or 70. The proposal was rejected. Other members, notably Simon Newcomb the astronomer, continued to call for governmental support. Aside from occasional governmental requests for advice, the Academy had no contacts with the federal government and received no funds from that source. It was an honorary body with no scientific operations under its wing. Not until World War I did the National Academy acquire an operating arm, the National Research Council, and become a greater force in the scientific community.

What was the significance of the founding of the National Academy of Sciences? Strangely enough, the most important aspect of the founding is that nothing happened. Most of the other great national academies became strong forces for research in their countries. But in America no powerful general scientific bodies came into being. Neither the Academy nor the AAAS ever dominated the scientific scene. Between the Civil War and World War I national organizations for specific scientific fields appeared. These soon controlled the publication of research results and the establishment of professional standards. The scientific community, in effect, turned its back on dreams of power in favor of tending its own gardens. The period 1865–1914 can be described as one of slow steady growth in science in America unaided by massive infusions of public funds and unplanned by any responsible bodies. Lack of funds pinched many scientists badly; the absence of planning resulted in odd patterns of strength and weakness—too many observatories and not enough physics laboratories to cite one example.

The defeat of the Lazzaroni by Henry and his allies prevented the stifling of science in America by confinement to an existing mold. For all their virtues, the Lazzaroni represented established science. They were receptive to innovations but, like everyone, only to a limited degree. Agassiz, who served biology in America well by stressing microscopy and physiology, castigated his opponents as "old fogies"; he never viewed his own opposition to evolution in that light. Gould and Peirce, the mathematical astronomers, underrated the Bonds. Unhampered by official standards, American scientists would roam widely, if not always wisely. Unconfined curiosity would lead many Americans into significant researches.

Agassiz to Bache

Cambridge March 6, 1863

My dear Young Chief,

Yes there is a National Academy of Sciences, and we may well rejoice. It inspires me to see how young you feel about it. I trust the Chiefess shares your enthusiasm, I am sure she does, judging from the impression I received during my last visit that she is truly one of us.

As soon as Wilson comes home I shall ask all our Scientific Men, which right or wrong, to meet him at my House.

Now let us proceed to organize in such a way, that our action shall bear the nearest scrutiny. I wish our first meeting would have some solemnity. It were best to gather for the first time in Philadelphia in some of the hallowed places of Revolutionary Memory. The learned *Grandson*[1] of Franklin must be our first President, and here shall the old man be pardoned for not introducing a clause in the Constitution favorable to Science, as he left a better *seed*.

Our first business should be to remedy the infirmity of the first appointments by submitting the whole again to a vote and making arrangements by which old fogies could be dropped from time to time, so that the Academy shall always be a live body. We ought to meet latest in May. How shall the first meeting be called. I wish it were not done by you that no one can say this is going to be a branch of the Coast Survey and the like.

Ever truly your friend. My love to Mrs. Bache, and from Mrs. Agassiz.

L. Agassiz

Rhees Collection, Huntington Library

[1] Bache was a great-grandson of Franklin.

Henry to Stephen Alexander[2]

Smithsonian Institution March 9th 1863

My Dear S.

I will endeavour to obtain for you a copy of the Bill establishing the National academy as soon as I can go to the government printing office.

It was carried through the two houses of congress by Mr. Wilson and the other member from Mass. at about 12 o'clock on the last night of the session without opposition. I had no hand in making out the list and indeed was not informed of the project until after the resolutions were in charge of Mr. Wilson.

I am not well pleased with the list or the manner in which it was made. It contains a number of names which ought not to be included and leaves out a number which ought to be found in it. The proper plan would have been to start with say two members[3] and to have given them an opportunity to fill up the remaining thirty by degrees after a thorough canvass of the several candidates. Instead of this the whole list of members to which the Society is limited is mentioned in the law of congress and therefore at present there is no room for the addition of other members.

I do not think that one or two individuals have a moral right to choose for the body of scientific men in this country who shall be the members of a National Academy and then by a political ruse, obtain the sanction of a law of Congress for the act.

The foregoing is my opinion of the affair but since the academy is now established by law either for good or for evil I think it becomes the friends of science in this country whose names are on the list to make an effort to give the association a proper direction and to remedy as far as possible the evils which may have been done.

A meeting has been appointed to take place as soon as the opinion of the members can be obtained as to the time of holding it in New York.

I think it probable it will be held in April or May.

All well, love to wife and little ones. Truly yours affectionately

Joseph Henry

Henry Papers, Smithsonian Institution

[2] Professor at Princeton and Henry's brother-in-law.

[3] Henry meant to write twenty. Did the slip of the pen indicate his distress that he and Bache were not working together and arranging affairs?

Bache to John Fries Frazer[4]

Washington March 10, 1863

Dear Grandson:

I am truly grieved to find you so out of health and spirits. How is it with you now?

There was very little about the Academy, except in spots. The act was easily drawn, and Senator Wilson offered to push it through. Then he wanted a list of names which was not easy to furnish and could have been however better by wider consultation had such been possible. Then the basis of qualification.

After looking at names of Washington members we concluded that it was better that Sen. Wilson should do up the Call for the meeting, aim to launching us. I went in for New York chiefly as the place of most *economical* meeting, seeing that that was likely to be an important consideration.

You and I hit upon the same time. I went in for the last of March or the first week in April.

No I do not think that Washington at this time is a good place for meeting. All the old associations have been *local* ones and we want to be universal.[5]

I send you a copy of the act of Incorporation, which gives us power to organize in our own way. All money clauses had to be left out, as requiring time to get before the Com. of the whole of the House of Reps . . .

Bache to Frazer

Washington March 12, 1863

Dear Grandson.

The Catalogue has not yet turned up, but it will no doubt, and at any rate it will be in time for I do not intend to recommend any body as against Dr. Frank Smith, and I am much obliged to you. It is no affair of mine and had not Van Buren asked me to do the thing I would not have thought of it, having a great aversion to recommending for subjects where my authority is and ought to be nil.

The gentlemen whom you handle so flawlessly under the title of

[4] Frazer (1812–72) was a professor of chemistry and physics at the University of Pennsylvania. He was not Bache's grandson; the salutation was a private joke.

[5] Bache is referring to the Columbian and National Institutes, earlier scientific associations in Washington with unsuccessful national pretensions.

Young Ben, is also down on the Academy list, though admitting that upon the whole it is unexceptionable.
I have been obliged to admit in reply that there are some men too mean to bring into our Academy thus slightly intimating that I so class Geo. P. Bond, and Spencer F. Baird. I have had favorable opportunities for inductions upon them in parts of their lives, and have come to distinct conclusions. The ven[erable] Smithson would I think have had both in and R. E. Rogers[6] out. That nomination seems to be *generally* looked upon with disfavor . . .

Frazer Papers, American Philosophical Society

Peirce to Bache

ϕ^r Grove 27_{III} '63[7]

Most Darling Chief:

I shall write to the Brevort House to day to secure a room for myself and *wife* on Tuesday April 21st. You can, therefore, if you wish write to them, if you wish to make any modification by which we may all have a parlour in common. How long shall we stay? Gray is already showing the cloven hoof. He is trying to divide Henry from us, as thus.

1. He asks Torrey "what does this mean about the National Academy."

Torrey knows nothing and is sure that Henry knows nothing because he was staying with him. He asks Henry about it and especially how the list was made out. Henry knows nothing of the list, till too late to amend it, and would have preferred that only fifteen or twenty should have been selected for the corporation, who should have elected the rest.

2. He asks Agassiz about the list and says that Henry says he was not consulted about the list. Agassiz says that he knows Henry was consulted, and took part in arranging it.

Now whatever be the true facts of the case, it is a nice pet to hang up a wet blanket upon.

Ha! Ha! Ha! The botanist may find that it is possible to dig too deep for successful undermining. His questioning of Agassiz, did not aid him much, I guess. But we must keep our eyes open; and never

[6] Rogers (1813–84), brother of W. B. Rogers, was on the faculty at Pennsylvania.

[7] B. Peirce is ϕ^r. 27_{III} '63 is March 27, 1863.

forget that Presbyterianism is a most mysterious form of concatination.

I had a letter to day from the Admiral [i.e. Charles Henry Davis] dated March 23 asking me if I would be willing to serve on the Compass Commission.

I shall answer "Yea Verily." God bless the dear chief and chiefess ø^ra[8] sends all love

your ever loving

ø^r

Benjamin Peirce Papers, Harvard Archives

Bache to Frazer

Washington April 7, 1863

Dear Grandson . . .

About the Academy I shall send you for remarks what appeared reasonable to Gibbs and me, if I ever can get time to put it in shape. If judged expedient we have all the power needed to make associate members it seems to me, but what you say in regard to them has great might. It forms a sort of grammar school, introductory to the College, and the best men might not accept. Alexander J.H.[9] has his mind clamped upon a rotary administration. My leanings are to a permanent one but of course under such suggestions as you make and others have made I must be reticent or incur a charge which would be unpleasant to me!

I like your idea of but two sections after much consideration. I was at first in favour of more, but now believe that two or three will better serve the cause.

Vice Presidents Henry, Peirce.

Foreign secretary Agassiz to avail ourselves of his universal acquaintances.

Domestic? Treasurer?

No permanent place of meeting, but where an impulse is wanted to science and calculated to do good.

Xmas and July, or Easter and Nov.

Grandma will go with me. If you want me to write for rooms let me know in time. . . .

[8] Mrs. B. Peirce is ø^ra.

[9] John H. Alexander (1812–67), a mathematician interested in metallurgy, weights and measures, and surveying. Bache originally tried to get the officers selected for lifetime tenure.

Look here, good Grandson, step in between me and any suggestion that I have been bound by ambitious views in the Academy matter, for there is no truth in the idea. Henry Wilson having said that he could get such an institution through I furnished him to the best of my ability what he asked for and then the S.I. meeting which nobody knows better than you was [two words unclear] from altogether different motives put Peirce, and Gould and Agassiz, etc. on the ground fortunately as I think, or the whole thing would have been badder.

Yours ever
ADB

Frazer Papers, American Philosophical Society

Henry to Gray

Smithsonian Institution April 15, 1863

My dear Dr.

I was glad to learn by your letter of the 10th that Dr. Torrey is on the mending bound. He left Washington in feeble condition and I presumed has been kept under by the bad weather of this backward spring. I trust however he will recruit on the opening of warm weather and hope he may long be spared to his family and friends. We can surely never see his like again.

I send you with this note a copy of the law of Congress establishing the National Academy, but can give you no information as to the plan of organization which will be proposed. I have made no enquiries in regard to it and have received no definite information.

The subject of an Academy was discussed about a month before the action of Congress by Davis, Bache and myself when we came to the conclusion that it would be impossible to obtain the passage of a law authorizing such an institution; and that if established it would give rise to so much bad feeling that it would be productive of little good. Instead of an Academy it was concluded to ask the appointment by the Navy Department of a Permanent Commission to which all questions of a scientific character should be referred.

This commission was appointed and has been in active operation for nearly two months. It has occupied nearly all my time, not devoted to this Institution and more than I could well spare. It has done good service and can scarcely be improved upon by the act of the Academy.

I shall attend the meeting of the Academy and do what I can to give it a proper direction. I put but little faith in appropriations of Congress. On the first application for an appropriation the friends of those who have been left out will make war upon the establishment.

I shall be most disappointed in not meeting you in New York since I have a number of subjects on which I wish to confer with you and which I cannot well commit to paper. Cannot we meet somewhere during the summer?

With kind regards to Mrs. Gray I remain truly your

Friend and servant
Joseph Henry

Gray Papers, Harvard

Leidy to Hayden

Philadelphia 1302 Filbert St.
April 28, 1863.

Dear Dr. Hayden,

Having been absent from the city, in attendance on the meeting of the corporators of the National Academy of Sciences I could not answer your letter of April 10 until now. I take the opportunity of saying I also received your former letter.

I did not remain until the end of the meeting, nor was I altogether pleased with its proceedings and doubt whether I shall retain my connection with the institution. It appears to me to be nothing more than the formation of an illiberal clique, based on Plymouth Rock.[10]

Joseph Leidy

Hayden Papers, National Archives

Agassiz to Bache

Cambridge May 23, 1863

My dear Bache,

I have to thank you for several papers which came a few days ago and for which I am very much obliged; but I write mainly to submit to you a circular which I have lately printed with the view of stimulating anew the efforts making in behalf of the Museum at Cambridge. The more I advance the more instinctively do I feel that if

[10] Leidy strongly objected to the requirement of an oath of loyalty advanced in the heat of the war and as a prerequisite for federal patronage. The oath was dropped in 1872.

I am permitted to work another ten years I may have brought this institution to the first position among the Zoological Collections. You can greatly help me now. I am about to arrange the Faunal Collections. I want specimens from every point along the whole Coast, with precise localities and the most common the object collected the more useful will they prove for this object. There is therefore no light house keeper and no tide observer who could not do real good service in collecting the materials to settle one of the most obvious points in our science, the geographical distributions of the animals along our Coasts, on the Atlantic as well as the Pacific side of the Continent. I write therefore to ask you whether you would let me have a complete list, with the direction of the light house keepers and tide observers all along our Coasts. Or if more convenient I would send you the requisite Number of copies of my circular to enclose in your official communication to them. But I would prefer the directions, as there are localities with reference to which I would like to have it in my power to make special remarks to the men of the station. I would like also to forward a batch of these circulars to every one of the Coast Survey stations.[11]

My dear Bache, I can not let this sheet go without expressing my delight at our success last month in New York. To have this organization settled is a great step, and I see the best fruits growing from it. The malcontents will be set aside or die out and the institution survive and it now remains for us to give it permanency by our own doings. It has already accomplished one great thing. We have a standard for scientific excellence, whatever our shortcomings may be. Hereafter a man will not pass for a Mathematician or a Geologist, etc. because an incompetent Board of Trustees or Corporation has given an appointment. He must be acknowledged as such by his peers, or aim at such an acknowledgement by his efforts and this aim must be the first aim of his prospects.

In my letter I write to Europe I have something about our National Academy. I should like to hear from you, whether you think it wise to have anything about it that may be read as officially coming from the Foreign Secretary, before our Rules are mentioned by Congress.

Ever truly your friend,

L. Agassiz

P.S. Give my love to the Chiefess.

[11] Agassiz was gathering data to use against Darwin.

Agassiz to Bache

Cambridge May 30, 1863

My dear Bache,

I have consulted the two Wymans[12] about a physician for your ship, the Coast Survey, who could at the same time do some good scientific work. But there is no one to be had, the best have gone to the Army and the indifferent ones are not fit to be encouraged.

I have all this time been thinking how I could make myself useful to the country in connection with the Academy. And as I have not been well since our meeting in New York I had full opportunity of thinking the matter over. Lately I was a few days on Cape Cod and upon seeing how people fish, it has occurred to me that it would be a good thing to have the whole subject of the fisheries considered from a scientific point of view. It would lead to very little expense and I believe be of great practical importance. I would propose besides me Dr. [Jeffries] Wyman and Dr. A. A. Gould be on that committee. We are all near together and could easily consult with one another. Should Dana be on it? I am not quite sure. I understand that he cannot work continuously and it would not be so easy to see him, to arrange plans. Though no one should be omitted who could render good service.

Many thanks for your very kind and affectionate note of the 29th, which is just received. I look forward with great expectations to the results that may be obtained from the execution of this plan of interesting light House keepers in Scientific researches.

I fully agree with you that we must make our Academy by our works, without asking for support or encouragements from anybody. I hope it may soon take in that way a position which will enable it to stand as a jury in the world, in matters of science. The idea of allowing the standing of Americans to be determined by European votes is perfectly odious to me. I hope on the contrary we may soon help them the better to appreciate their own men. Example: This very year the discoveries of Boucher de Perthes[13] are for the first

[12] Jeffries Wyman (1814–74) was a comparative anatomist at Harvard. His brother Morrill Wyman (1812–1903) was a physician.

[13] The French archaeologist Boucher de Perthes (1788–1868) did notable research on prehistoric man. Gliddon and Nott were Americans whose book had argued that the different races of men were not of the same species, a view Agassiz supported. The South applauded and orthodox Christians and humanitarians were outraged. But in a few years Agassiz' attacks on Darwin had restored him to favor among many religious circles.

time discussed in the French Academy. You will find them analysed in a book on the Races of Man by Gliddon and Nott, published Philadelphia nine years ago; and for years I have been teaching the doctrine of a succession of human races, different from one another, in somewhat the same way as in earlier geological periods, animals of different kinds often followed one another. But this is not yet looked upon as even a possibility. Nous verrons.

Ever truly yours,
L. Agassiz

P.S. I perceive that Dr. Wyman nor Dr. Gould nor Prof. Dana have as yet qualified themselves to act as members of the Academy so that I am the only Zoologist who could be at once appointed. I would be sorry to have this matter postponed to next year, as there is one spawning season before us, which ought not to be lost.

L. A.

Rhees Collection, Huntington Library

Leidy to Hayden

Philadelphia June 7th 1863

Dear Dr. Hayden

. . . I can't send you a list of the "eminent savans" as you desire for I have not one myself. I think it will turnout to be a grand humbug, and I intend having nothing to do with it. A society of the kind that leaves out such men as Baird, Draper, Hammond, Lea, Cassin, yourself,[14] and appropriates a number who never turned a pen or did a thing for science, certainly can't be of much value. . . .

Joseph Leidy.

Hayden Papers, National Archives

Henry to Agassiz

Smithsonian Institution Aug. 13th 1864

My Dear Professor

I have just returned to this city and find your letter of the 8th awaiting my arrival. At the Depot at Philadelphia I met our friend Mr. Felton and after some remarks relative to the family of his

[14] Hammond was then still Surgeon General. Isaac Lea (1792–1886) was a Philadelphian who was an authority on mollusks. John Cassin (1813–69), another Philadelphian, was an ornithologist.

lamented brother;[15] turned the conversation to yourself; and said, that when he saw you, a short time before, he thought you were looking very ill, that you were too much occupied with various matters, and that he had strongly urged you, on your own account, and on that of your family, to give up all care, and for a time to think of nothing but the reestablishment of your health.

The perusal of your letter has rendered the importance of the advice of Mr. Felton strikingly evident to me; and in view of the present condition of our much esteemed friend Professor Bache, whose malady, I trust may be but temporary, I beg that you will take a more cheerful view of the proceedings at New Haven; or rather that you will banish them entirely from your mind. It is of much more importance to the science of the world that your health and life should be preserved than that the academy should be rapidly advanced to your ideal standard of perfection. In this Democratic Country we must do what we can, when we cannot do what we would. We must expect to be thwarted in many of our plans and learn to bow before defeat with the consolation of knowing that if we have not succeeded in our aim we have at least deserved success.

After a calm review of the proceedings at New Haven I think them much more favorable than under all the circumstances of the case there was reason to expect they would be at the commencement of meeting.

Permit me to give you a candid and free exposition of my views of the matter and for this purpose to go back to the beginning of the academy. Several weeks before you and the other originators of the academy came to Washington Professor Bache asked my opinion as to the policy of organizing a National Association under an act of Congress. I stated, in reply; *First* that I did not think it possible that such an act could be passed with free discussion in the House, that it would be opposed as something at variance with our democratic institutions. *Second* that if adopted it would be a source of continued jealousy and bad feeling an object of attack on the part of those who were left out. *Thirdly,* that although it might be of some importance to the Government yet it would be impossible to obtain appropriations to defray the necessary expenses of the meetings and of the publication of the transactions. *Fourthly* that there would be great danger of its being perverted to the advancement of

[15] C. C. Felton (1807–62) a classicist who was one of the rare non-scientists among the Lazzaroni. As president of Harvard (1860–62), Felton backed Peirce and Agassiz in that institution's intramural warfare.

personal interest or to the support of partizan politics. With these views, I thought, Professor Bache was impressed. He said no more to me on the Subject and I heard nothing further in regard to it until after the whole scheme was organized and in charge of Mr. Wilson of the Senate.

Besides the objections I had presented to Professor Bache I did not approve of the method which was adopted in filling the list of members. It gave the choice to three or four persons who could not be otherwise than influenced by personal feelings at least in some degree; and who could not possibly escape the charge of being thus influenced. I did not however make any very strenuous objections to the plan because I did not believe it could possibly become a law; and indeed there are very few occasions when acts of this kind could be passed without comment or opposition. After however it had become a law I resolved to give the academy my hearty support, and I have since faithfully and industriously endeavored to advance its interest.

My anticipations in several particulars have been realized—and antagonism, such as I feared, has been produced in the minds of those who think themselves ill used in being left out; while a considerable number of those who were elected feel that they ought to have been consulted in making up the list of names. The feeling also exists, to a considerable extent, that the few who organized the academy intend to govern it; and I think that this was the *animous* which excited the determination to elect Professor Baird. He was the choice of a large majority of the cultivators of Natural History; and although your opposition was honest in intention, and your position correct in general principle yet I think that had you prevailed in your opposition, a majority of all the naturalists would have resigned; and a condition of affairs would have been produced deeply to be deplored. I fully agree with you in opinion, and I presume the philosophical world would also concur with you, that as a class of investigations those which relate to Phisiology and the mode of production and existence of organic forms are of a higher order than those which belong to descriptive Natural History. The good however which two persons may have done to science in these two classes will depend on the relative amount, as well as, on the character of their labours. Besides this you ought to have commenced with the application of the principle of the higher investigation, in the formation of the Academy, for you could not reasonably expect that any member would vote to disparage his own pursuit.

It is true I voted for Mr. Baird and taking all things into consideration I am sure I did right in doing so. I do not agree with you in thinking that my having voted for him will give him the power to control the policy of the Institution; neither do I think that the proposition you made at the meeting of the Board of Regents has any connection whatever with the vote in the Academy. It is the same which I have advocated from the first, and which I doubt not will meet the approval of the majority of the intelligent naturalists of the world. If Mr. Baird would attempt to interfere with the policy of the Institution I would not hesitate to ask him to resign, and to insist on his doing so; as I did in the case of Mr. Jewett. But I have not the least idea of any trouble with him in this way; or that for many years to come any thing (indeed, in the way of carrying out your proposition). Had the war been brought to an end last spring we might have indulged a hope of this kind; but the injurious additions which have been since made to the national debt will induce a very cautious policy in regard to the application of the public funds.

You do me but simple justice in supposing that I would not willingly join in any intrigue to advance personal or family ends. In the whole course of my life I have never engaged in anything of the kind and it is now too late for me to change my character in this respect. It is necessary however sometimes to have an eye on the acts of others in order to thwart their improper designs.

I think you are regarding this matter of the Academy in so serious a light that it will unfavorably affect your health and spirits. I fully agree with you in opinion as to the desirableness of elevating the standard of American science; but we must recollect that great changes are seldom or never produced *percaltum* and that we frequently waste our strength in endeavouring to suddenly overcome an obstacle which will gradually give way under a gentle constant pressure. I fear had you succeeded in excluding Baird from the Academy on the ground of the character of his investigations you would have aroused a large amount of personal opposition and have been subjected to criticisms and other annoyances which to a nature like yours, craving love and sympathy, would have been exceedingly painful. Why trouble yourself so much about the character of American science which can only be improved with the social and political conditions which tend to encourage and develop it. You have already done good service by your presence in this country, by your immediate instruction and by the enthusiasm and sympathy which you never fail to awaken. You are formed to lead men by the silken cords of love rather than to urge them on by the rough method of coercion. Let me beg of you therefore, my dear Professor, to first take care of your health and secondly to devote yourself for the remainder of your life to those investigations which have given you so wide a reputation and in which at every step you can elevate yourself in your own self esteem as well as in the admiration of the world, afford to look down with complacency on the means to which ordinary men resort to raise themselves into temporary notoriety.

It is lamentable to think how much time, mental activity, and bodily strength have been expended among us during the last ten years, in personal altercations which might have been devoted to the discovery of new truths; to the enlargement of the bounds of knowledge, and the advancement of happiness. There is a cause for this which might be discovered; and I will venture just to mention a principle of action which may have had some influence in producing the results. I allude to the principle of supporting our friends right or wrong. I grant that this principle of action springs from a generous impulse of a warm heart; but it does not receive the approval of the moral judgement of a cool head. We are not true to our friend if we follow, or assist him, in a single step in the wrong direction. You lost an invaluable friend in Professor Felton who with his unselfish disposition, and expanded sympathy ever acted as a bond of harmony and union between all the varied characters that constitute the faculty and associates of Cambridge; and should it be the design of Providence to remove Professor Bache, which Heaven in its mercy forbid, a similar change, though of a some what different character will be felt in the circle of which he is the controlling center. . . .[16]

Benjamin Peirce Papers, Harvard Archives

Henry to Bache

Washington Aug 15th 1864

My Dear Professor

I reached home on Wednesday in a tremendous hot term, in which there has been scarcely no let up of any consequence for several weeks. I came in the Cars with Henry Carey[17] who made kind in-

[16] Note of Aug. 28, 1864, omitted.

[17] Carey (1793–1879) was notable for his advocacy of protectionism in a period when free trade was intellectually fashionable.

quiries for you, and amused me very much with his very ingenious and novel, though partial views in Political economy.

I found wife and daughters well, but almost melted with the extreme heat. The vegetation of the Smithsonian grounds are as scorched as if a fire had passed over it and every where the indications of an extreme drought is more striking than I ever before beheld them.

The meeting of the Academy went off on the whole, very well, although our friends, from Cambridge, were much displeased because they did not succeed in preventing the election of Professor Baird. They were right in principle but wrong in practice. I fully agree with Agassiz that physiological research is a higher order of scientific investigation than the description of species but the *amount,* of labour as well as the *kind,* must be taken into consideration and as all the naturalists with the exception of Agassiz and one or two others have made their reputations and were elected into the Academy on account of just such work as Baird has done the opposition of Agassiz who stood alone was considered against the whole class and had he recommended in keeping Baird out by urging a rule of the Academy, the majority of naturalists would have withdrawn.

It will not do to press a principle in some areas too far. In this democratic country we cannot expect in all instances to carry our points and must be content in doing what we can instead of what we would.

I visited Capitol Hill on Saturday evening and had considerable conversation with Schott and Saxton who are both loyal and true subjects of yours. They both informed me that all the operations of the Survey are going on smoothly and that thanks to your admirable organization they may continue to do so without your aid until next summer.

I did not see Hilgard who was absent at the time. Indeed I called on Saxton and Schott [18] after the offices were closed and saw them at their own houses. . . .

[18] C. A. Schott (1826–1901), Joseph Saxton (1799–1873), and J. E. Hilgard (1825–90) were Bache's subordinates at the Coast Survey, the last two being members of the Academy. Schott worked for the Survey for almost forty-five years as their principal mathematician. Saxton was in charge of the weights and measures; his specialty was precision instrumentation; Hilgard was a very competent geophysicist who was second-in-command to Bache.

Henry to Bache

Shelter Island Sept. 9th 1864

My Dear Professor

I am rejoiced to learn from your letter of the 2nd that you are rapidly improving and I trust that you will give up all thoughts of business and attend exclusively to the care of your health looking cheerfully hopefully on the sunny side of the future.

All the establishments in which you are especially interested are in a prosperous condition. The Coast Survey is going on well and with the admirable system of organization to which you have reduced it will continue to go on in this way for months to come by the mere influence of your name, and the power of your will, though the latter may not be at present actively exerted.

The Smithsonian is in its usual condition, and I apprehend no great difficulty in the future unless Agassiz should attempt to carry out, at an improper time, his proposed scheme of improvement.[19]

The Light-House Board with Harwood, who is one of your most judicious and warmest friends, as the secretary will fully cooperate with the Coast Survey.[20]

The National Academy with the judicious direction which you can give it will, I doubt not, become an important aid to the government, and a potent means of advancing American Science. I say with your judicious direction; for without this it will soon fall into confusion if our friends at Cambridge attempt of themselves to control its operations. Without you they can have no power.

You must take care of yourself in the present; for in the changes which our country is destined to undergo you will probably be wanted in the adjustments of difficulties of a material character even more important than those which yet have been introduced into your care.

I will have a free and full talk with you on the affairs of the academy when we meet. The late meeting terminated much more favorably than could have been anticipated at the beginning.

I would have visited your camp this week had I been assured that you had returned; but owing to the bad weather I have remained

[19] Probably a reference to Agassiz' attempts to get the National Museum Collections transferred to his museum in Cambridge.

[20] Commodore A. A. Harwood (1802–84). The Light-House Board was a Henry-Bache venture in co-operation with the military. Henry carried out experiments for the Board on light and sound transmission.

with Professor [Eben N.] Horsford [21] longer than I intended. He is truly a very amiable man, and a warm friend of yours. I have never known a man who thinks more kindly of all his acquaintances.

I have written a long letter to Agassiz urging upon him, to devote the remainder of his life to the investigations which have made him so widely and so favourably known; and not to trouble himself about the condition of American science which he can better advance by his example in the way of actual work, than by coercion. In the last twelve years an immense amount of time and feeling has been wasted in quarreling, this should not be and I hope when you get well you will join me in attempting to put an end to this state of things.

We must endeavour to control Gould; and now that he is married to get him to see that the world is not as bad as he has thought it to be.

Dr. Torrey, Guyot, Alexander of Princeton, and many other members of the Academy are true men on whom you may always depend to do what is just and proper but they have said that they would rather leave the Academy than be continually subjected to the annoyances of disputes as to the policy and government of the establishment. I hope however that all will go well, and I shall not fear for the result provided you take care of yourself and be prepared at the proper time to take the direction of affairs.

Mrs. Henry and Helen are still at Chestnut Hill, and I shall return there in order to accompany the latter on the excursion I have promised her. We arrived here on Saturday last. Mary and Caroline will remain a week or ten days longer while I go to the West with Helen.

Direct your next to me care of Dr. Torrey Assay Office N.Y.

As ever truly yours
J.H.

Henry Papers, Smithsonian Institution

[21] Horsford (1818–93) was a chemist who was at Harvard until 1863. He was involved in research with the Permanent Commission and the National Academy.

Leo Lesquereux to J. P. Lesley;[22] excerpt

Jan. 21–30, 1865

. . . I thank you indeed most heartily for your open and free manner of speaking to me about what you heard concerning Gray, Agassiz and others. I suppose that I knew Agassiz very well and had seen him through [one word blotted out]. I was far from having an idea of such a violence of character a[s] you represent to me from his own words. Of what he said I wrote him about Gray, this only is true: He, Agassiz, wrote me from New Haven the same day of my nomination to inform me of it and he did it in the most flattering terms. He mentioned nevertheless that Gray had opposed my nomination and would not hear to have me in his section and that when he saw that you wanted me with you, he (Gray) had immediately changed practice and tried hard to have me nominated in the Botanical section. To this I answer: that I was indeed surprised of this because Prof. Gray had always been a good and kind friend to me but that of late I had observed some change in the manner of his acting (*procédès*) towards me and that I could not explain the change. It is all what I have written about this matter. I have supposed and I believe still that Gray was not pleased at my publishing the California mosses without Sullivant [23] and of my publishing them in Philadelphia for he does not like your Proceedings and Transactions. Though he reported with eulogium Mitten's mosses published in the Linnean of London he did not say one word of my California mosses which I consider as a far better work except perhaps for descriptions, than Mitten's. As to me that time he stopped sending to me copies of his periodical publications, I considered this as a change of opinion against me and it is from this persuasion that I wrote to Agassiz the few words mentioned above. Sullivant has done very much for Gray and I do not wonder that Gray loves him. But Sullivant is also a good friend of mine though I owe him nothing and I have worked hard and gratuitously for him. He had not time to work either with me or by himself. . . . Lately Gray has written

[22] Lesquereux (1806–89), a paleobotanist, came to America at Agassiz' behest. Lesley (1819–1903), a geologist who later headed the geological survey of Pennsylvania. Lesquereux was elected to the National Academy in 1864, apparently without the fireworks attending Baird's candidacy but over Gray's opposition.

[23] William Starling Sullivant (1807–73), a botanist specializing in the mosses.

a good letter to me and sent me also copies of his pamphlets. Hence I may have been mistaken in my impression or he may have changed his mind. As I do not care a *pip* about the opinion of those high gentlemen and appreciate my independence far more than any thing they could do for me, it did not trouble me much about what was written or said to me on these matters. But what pains me much indeed is the violent hatred and jarring despite openly proclaimed by men of high standing whose influence should be only good and productive of a higher tone of moral and gentleness. In such disputes as those of which you speak to me, there is no trace of dignity, not even of decency. Such men would sacrifice to their little but furious passions an institution which should become the most honorable for a country and if not potent enough to do it through disrepute and contempt on matters which should be considered only with respect even with veneration. Yes! the subject has pained me much. But what can we do. We expect more of those we call great men and they are but men after all . . .

J. P. Lesley Papers, American Philosophical Society

A. Guyot to Henry

Princeton N. J. February 13th 1865

My dear Sir,

On my return to Princeton, after the Winter vacation, I learned, with real grief, the sad and irreparable loss you have sustained by the fire of the Smithsonian building, of all your records of original observations. This is more than a simple personal loss, which adds to the gravity of that national calamity.

I had read in the papers the news of that unfortunate fire, but I little realized that so many valuable papers could not have been saved from destruction. Though the sympathy of your friends, however deep, cannot restore to you those precious documents, I feel constrained to express mine most heartily. I thank God that none of your family was hurt. My sister and myself send them all our most affectionate regards.

I had the pleasure of having Agassiz spend with me Sunday 15th Jan. I touched upon the tender question of his relations with Gray, by assuring him that having been present, at Dana's house, that all that had been said, and had transpired about Gray's and Dana's action in the election so much contested by Agassiz and others I could declare that not any other feeling had been expressed, or harbored by any, but that of preventing what appeared to all an unjust treatment of the candidate; and so it was.

Now that the death of the lamented Gilliss has opened the door of the Academy once more, it seems to me that Bond should be called in.[24] Our Cambridge friends had their own way in the last meeting. Let justice have its own in the next.

I remain, my dear sir, with great affectionate regard

truly yours

A. Guyot

Henry Papers, Smithsonian Institution

Lesquereux to Lesley, excerpt

March 21 1866

. . . I am satisfied that Gray will not give his esteem and friendship to one who does not merit it. Gray is somewhat autocratic of character like Agassiz. He well knows that he is a prince of science and if he does not openly dispise small men and poor things, he keeps his whole regards for high standing subjects. But he is certainly frank and honest and dispises every kind of duplicity. I see it in that way at least. . . .

I wish I could have a long talk with you about the A. Academy. As it is constituted now it appears to me a kind of anomaly considered with the eye of the American character. Nothing is good or productive of good in this country if it is not either sustained by money or bringing money to the owner. What will you do with a body of scientific men who, in *America!* are expected to work for nothing and what influence can have a scientific society, the so called highest scientific society of the Land, which cannot even publish a scientific memoir. In Europe, honor conferred is worth more than money but in America the same honor is worth nothing by itself if the Academy or any other Scientific Association does not find the lever for showing its worth, viz. money you may augment the number of members per thousand without any appreciable result. The want of a scientific body as a nucleus to which should converge all the rays of peculiar or private scientific societies of the Country is to my mind evident. But without the means of either attracting or throwing out these rays of light, the body is a mere rough diamond whose value is ignored and thus nearly useless. If the number of the

[24] Bond died before he could be elected.

members shall be augmented I would say: 1st Each member of the National Academy is willing to pay $25 annually for the expenses of publicatio[n]. 2nd Among the members which aspire to become members, the first consideration of election should be personal scientific value; the second willingness of submitting to the same amount of taxation for publication of memoirs. I would rather pay $25 annually and read the memoirs than see them put aside without any use. . . .

Lesley Papers, American Philosophical Society

Gray to Henry

April 11, 1874

My Dear Sir

Permit me to make a suggestion to you, as the President of the National Academy of Sciences.

I am led to think that the present is a most favorable time to endeavor to initiate a certain change in the organization of that establishment, which would practically ensure for it a healthy and prosperous existence, under the best wishes of all.

Merely to add the word *Washington* to its title, which gives a *local habitation* while it sufficiently indicates what *Nation* the Academy belongs to.

I think the present members from a great distance, not being in the service of the U. S. would be as well pleased to be *associate* members, and that the Academy taking root, and being a proper *raison d'etre* would begin to flourish and in due time be much more important than its elder sisters, the American Philosophical Society, and the American Academy of Arts and Sciences.[25]

While President of the latter I could not for a moment entertain the idea of any *perrogative* of the Washington Society such as its original organization seemed to claim. But I fancy that experience has shown that the plan adopted has a serious practical disadvantage.

Whatever you may think of my suggestions I am confident that you will take no umbrage at my having offered them, for what they are worth.

Very truly yours
Asa Gray

[25] Gray was a stalwart of the American Academy in Boston. Unlike Bache, his experiences with the federal government had not been very happy, and Gray favored localism rather than strong national groups.

Gray to Henry

Feb. 21st, 1877

My dear old Friend

Yours of the 15th ultimate is received, and has been read with much interest.

It seems that you may take courage in the hope of seeing your views and wishes for the Institution carried into effect.[26]

The initiation by Congress itself of action which would result in the separation of the National Museum from the Institution, is what, as you know, I regard as the best thing of all, but I hardly expected it. The affixing by Congress of the condition you referred to, viz. that the building shall be by itself, is the first and most important step.

I am interested to see, too, that the objection you felt at having the Smithsonian edifice left on your hands, may also disappear. The income you might derive from having the signal-service bureau and I will add the Light-house Board, as tenants (both particularly proper tenants), would make up for any loss. "Keep a thing a dozen years and you will find a use for it." says an old proverb. The signal service may utilize your *towers,* which I rarely pass without a mental gnashing of the teeth at the amount of money gone up into the empty air. But there is one thing to consider in case Senator Morrill's views are embodied in the appropriation. The sort of building, planned by Gen. Meigs[27] was appropriate for its purpose and its position, but would be inappropriate for the National Museum by itself, both for practical use, and as an edifice to stand by itself in the grounds. For abundant reasons, which I need not refer to, it is clear to me that you will go to a different building, one with *foundations* and a cellar,

[26] Correspondence between two old friends discussing how to stop the tides from coming in. Henry and Gray believed still that by making the National Museum a separately housed federal bureau, the Smithsonian would avoid burial in its collections. Henry was still trying to get Owen's white elephant, the multi-towered building, off his hands. Henry's strategy was to get rid of programs after a certain point but he always remained tethered to white elephants. Bache knew instinctively that the correct strategy was to acquire and to keep programs. With the Academy impotent, only the Smithsonian could speak for science in America. Henry tried manfully but his failing powers were largely committed to such schemes as renting space in his building.

[27] Sen. Justin Smith Morrill of Vermont (1810–98) is best known for his Land-Grant College Act. M. C. Meigs (1816–92), Quartermaster-General, was an Army engineer and a member of the National Academy.

and two stories. In fact you will probably come to a quadrangle, of which you will now build one or two sides.

We remembered, last evening, the reception at your house, and long to be there, mingling with the throng.

I am going in an hour to Admiral Davis' funeral, to act as one of the pall-bearers, four civilians to join the four officers.

Ever sincerely yours

Asa Gray

CHARLES SANDERS PEIRCE: THE PERILS OF GENIUS

To refute European charges of the neglect of theoretical sciences, knowledgeable Americans in 1900 could and did point to the great mathematical physicist at Yale, J. Willard Gibbs (see below, p. 315–22). They could also have pointed to Charles Sanders Peirce (1839–1914) but did not, for the simple reason that very few had heard of him. And those who did know of Peirce probably did not realize the extent of his contributions to science; these became evident only with the posthumous publication of Peirce's unpublished writings.

Peirce was the unexpected, exotic culmination of the geophysical tradition in America. In the home of his father, Benjamin Peirce, he grew up in an atmosphere where the intricacies of mathematics, astronomy, and geophysics were taken for granted. At an early age he went to work for the Coast Survey, later headed by his father. Here Peirce had an honorable position with a satisfactory income, congenial scientific work, and an outlet for some of his scientific writings. Peirce eventually became the Survey's specialist on the determination of the value of the force of gravity, swinging his pendulums in Europe and America. At one time he was in charge of the weights and measures, the precursor of the National Bureau of Standards. The Survey treated Peirce very decently, allowing him to live in Cambridge and giving him ample freedom to prosecute his varied intellectual interests. But the tradition which nurtured the agency was losing its force, and the successors of Alexander Dallas Bache and Benjamin Peirce were lesser men. When their administration became lax, a wrathful, economy-minded Congress and a Superintendent of the Survey who saw no merit in Peirce's work forced his resignation. He retired in 1891 after thirty years' service. The federal government did not have a pension system for its civil employees then, and Peirce was left without financial resources of any consequence.

There was no place at an American university for a distinguished philosopher-logician-mathematician-astronomer-geophysicist. He had alienated Charles W. Eliot, the President of Harvard, where his father and brother taught. Before his dismissal from the Survey he had been a part-time instructor in logic from 1879 to 1884 at the newly established Johns Hopkins University in Baltimore. Besides not attracting many students, Peirce became embroiled in a controversy over priority with a fellow faculty member, the great English mathematician James Joseph Sylvester (1814–97). His divorce and remarriage did not improve the chances for an academic post.

For the remainder of his life after leaving the Survey, Peirce subsisted on occasional writing jobs, lectures, and the charity of friends, notably the philosopher William James (1842–1910). During these years Peirce devised many schemes to support the completion of his projected philosophic masterpiece. It was forever in process, remaining incomplete at his death. To his contemporaries who knew of the work, the projected work was probably an unsuccessful grandiose scheme of a brilliant man gone astray. Peirce simply did not fit into any conventional pigeonhole. In an empirical age and place, he was a rationalist and a theorist. At a time when specialization was taking command, Peirce remained uncomfortably universal. Nineteenth-century America had no place in its scheme of things for a mathematical-logician and philosopher of science.

At first Peirce was remembered as a founder of pragmatism, and his views were confused with those of William James and John Dewey. After several decades of research it is recognized that there is much more to Peirce than that. While all three men shared an interest in the psychological aspects of thought processes, Peirce stressed the role of theory building and of how data is defined by where it is fitted into a theoretical structure. Part of the difference between Peirce and the true pragmatists is certainly ascribable to differences in personality and intellectual style. But much of the difference derives from Peirce's background—work in mathematical logic and in very precise physical experimentation and observation.

Benjamin Peirce to Mrs. L. R. Peirce

Cambridge Sept. 10 '39

Good news! Dearest mother, Sarah is lovely and so is the, yet anonymous boy. Sarah sends her very best love. She has had the least bad time she ever had, the least exhausting. She suffered for about two hours and at 12 was confined. The boy weighs 8¾ pounds, and is as hearty as possible. He offered to go out and throw the vase with me today, a certain sign that he will make an eminent lawyer or a thief. As soon as he publishes his "Celestial Mechanics" I will send you a copy, and I have no doubt he would be glad to correspond with you about your last mathematical researches. He has two splendid optical instruments each of a single achromatic lense which is capable of adjustment! And so wonderful is the contrivance for adjustment that, by a mere act of will, he is able to adapt it to any diam and at pleasure. But one fault has been found with these instruments and that is, the images are inverted; our new born philosopher, however, our male Minerva, contends that this is a great advantage in the present topsy-turvy state of the world. By the way, he calls himself Minervus. The first proof of his genius which he exhibited to the world consisted in sounding, most lustily, a wonderful acoustical instrument whose tones, in noise and discordancy, were not unlike those of fame's fish-horn. Is not this a singular coincidence? A sure omen of his coming, almost come, celebrity? But good bye and don't forget

that sad dog
but glad father
and loving son Ben

Benjamin Peirce Papers, Harvard Archives

C. S. Peirce to William James

Paris 1875 Nov 21.

My dear Willie

Your letter led me to look up your brother[1] whose presence here is a great thing for me as I am lonely and excessively depressed.

Your notice of Wright[2] is good. As to his being *obscure* and all that, he was as well known as a philosopher need desire. It is only when a philosopher has something very elementary to say that he seeks the great public or the great public him. And as for Cambridge being a village, it is no doubt but the only reason is that it doesn't believe in the possibility of any great advances in science or philosophy being made there and thinks the highest thing it can be is a school. A place may be far more out of the world and yet be as good a centre for philosophy as London or Berlin. As for Berlin, I really don't see that it has much claim to be thought a place for clear thought. The Germans always muddle whatever can be muddled. You are very kind in wishing me back in Cambridge. I don't know whether I shall ever live there or not. I like the place but there is something about it too which I find very antagonistic to me. I had ambition once to be a Professor there, which I have outgrown. Why put myself in such a position of obloquy almost if I could, why be

[1] Henry James, the novelist.

[2] Chauncey Wright (1830–75) was a mathematician in the National Almanac Office in Cambridge. He greatly influenced Peirce, James, and Oliver Wendell Holmes. Wright, like Asa Gray, defended the neutrality of science in the evolution dispute.

Charlie Eliot's man when I have already a position where I am engaged in original research and where even that is my *duty* and is counted positively for me, instead of being something to be excused as it is College. As for the observatory, it is of all situations I know of the one which has the most of thankless, utterly mechanical drudgery, together with vexatious interference from two different sources, certain members of the committee and the president. I speak of *Directorship* of it, for my own connection with it was most delightful. Winlock[3] was charming. I am not quite through with it because my book of photometric researches isn't out.
And they began at once after Winlock's death to try to nag at me, but fortunately I am too far off.

But don't let me speak as if I did not feel the warmest gratitude to you and my friends who wanted to get me into the observatory. I don't know that I would have declined it even although it does not seem to me altogether desirable. But I always speak too strongly when I think of Eliot. [C. S. Peirce's insertion]

I wish I was in Cambridge for one thing, I should like to have some talks about Wright and about his ideas and see if one could not get up a memorial of him. His memory deserves it for he did a great deal for every one of us. I don't speak of Warner and such philosophical *canaille* but I mean you, Frank Abbot, and myself. Other of his friends, Gurney, Norton, Peter Leslie, Asa Gray etc. would be wanted to do the personal and other relations. But what I am thinking of (I don't *purpose* any thing) is to give some resume of his ideas and of the history of his thought.[4]

Your brother is looking pretty well, but looks a little serious. He is a fine fellow. I have always thought I should admire him if I knew him better and now I shall find out.

Yours affectionately
C. S. Peirce

[3] Joseph Winlock (1826–75) succeeded G. P. Bond as head of the Harvard Observatory.

[4] In 1877, Charles E. Norton (1827–1908) issued Wright's *Philosophic Discussions* with a biographical memoir. F. E. Abbot (1836–1903) was a minister who engaged in the popular learned industry of reconciling science and religion. E. W. Gurney (1829–86) was an historian at Harvard. Peter Lesley (1819–90), the geologist, was another friend of Wright's. Joseph Warner (1848–1923), a lawyer, participated in the discussions at Cambridge involving the founders of pragmatism.

C. S. Peirce to William James

Paris 1875 Dec. 16

My dear Willie

I hear from my father that you have written a beautiful letter to the President of the Baltimore University[5] proposing me for the chair of logic and I am asked if I would accept.

It is a question impossible to answer in my present state of information. I don't know what the conditions are.

My place in the Coast Survey is agreeable to me. I have with great trouble learned to swing pendulums and Uncle Sam has spent a good deal of money upon teaching me. It is a very difficult business. I do not know what I could honorably disconnect myself from the Survey now. Just at present, I don't *think* I could.

On the other hand, I feel that in a Chair of Logic, I should be using my best powers, which I never shall use elsewhere. Elsewhere, in an observatory or anywhere, I shall always be a mediocrity, which does not disturb me in the least. In a Chair of Logic, I shall reach some eminence and leave some ideas which would ultimately be found very useful, I don't doubt. It is not very clear to me what I ought to do. But if I don't change my profession very soon, I never shall.

On the whole, I think I will leave the question to my wife to decide as she is on the ground.

I see your brother very frequently. He is a splendid fellow. I admire him greatly and have only discovered two faults in him. One is that [h]is digestion isn't quite that of an ostrich and the other is that he isn't as fond of turning over questions as I am but likes to settle them and have done with them. A manly trait too but not a philosophic one. He is looking better than when he first came and Paris is the place for him; Paris and he are adapted to one another.

Please remember me with great respect and love to your father.

Yours very affectionately
C. S. Peirce

William James Papers, Harvard University

[5] Daniel Coit Gilman, whose Johns Hopkins University marks the true start of graduate education in the United States.

C. S. Peirce to G. F. Becker[6]

Milford Pa. 1893 June 11
(12 M. Ther. in study 77°)

My dear Ferdinand:

I gave a great deal of serious thought to your paper, and then, later, came a rush of business, and it slipped my mind.

I admit that it is quite an important improvement upon the theory of elasticity, and that it seems to satisfy existing observations.

Turn the matter as I will, I do not see how any evidence whatever can prove, or even render likely, the exactitude of the expression of any continuous law whatever. It certainly cannot do so upon *my* theory of probable inference, nor can I see that the thing is possible upon any rational theory of inductive reasoning. I hold that no laws are exactly *obeyed;* but I do not deny that, there *are* real laws. But I ascertain *precisely* what those laws are, even if we assume that they have any *precision* in what they *require* (obeyed or not), that seems to me clearly impossible.

Gravitation varies inversely as the square of the distance; but how can we know that it may not be as the −2.000000000001 or the −1.999999999 power?

Now, when we come to molecules, what is the résumé of all we can with any probability guess about them? It certainly is that the laws of their attractions are of a complicated character. Then how can it possibly be supposed that so simple a law as yours, or Hook's, should be exact? It is true that it fulfills certain extreme conditions which the true law cannot depart much from. Nevertheless, it does not seem to me improbable, from all we know of molecules, that it does depart tolerably far in middle distances. In fact, I can hardly think that all substances are likely to have precisely the same elastic law.

Science has hitherto been proceeding without the guidance of any rational theory of logic, and has certainly made good progress. It is like a computer who is pursuing some method of arithmetical approximation. Even if he occasionaly makes mistakes in his ciphering, yet if the process is a good one they will rectify themselves. But then he would approximate much more rapidly if he did not commit these errors; and in my opinion, the time has come when

[6] George Ferdinand Becker (1847–1919) was an old friend of Peirce's from Cambridge. He was a geologist concerned notably with the applications of physics and mathematics to geological processes.

science ought to be provided with a logic. My theory satisfies me; I can see no flaw in it.

According to that theory universality, necessity, exactitude, in the *absolute* sense of these words, are unattainable by us, and do not exist in nature. There is an ideal law to which nature approximates; but to express it would require an endless series of modifications, like the decimals expressing surd. Only when you have asked a question in so crude a shape that continuity is not involved, is a perfectly true answer attainable. I return the MS. very faithfully

C. S. Peirce

C. S. Peirce to Becker

84 Broad St. New York 1897 Jan 14

My dear Ferdinand:

I have today received a request from the Director of the Geological Survey to express an opinion about your views on Slaty Cleavage and have replied asking him to send me the arguments on both sides. Of course I read your memoir on the subject and am perfectly aware that you are right. Darn not so clear that the other party is not right too in a measure. That is, not being a geometer, he has no clear ideas as to the direction of pressure. He probably thinks that if a pasty mass is forced between two parallel walls the pressure is normal to those walls, and that if one plane moves down perpendicularly toward a parallel plane, thus squeezing out of shape a mass between, the pressure is in the direction of the motion of that plane. Clearly, the planes of cleavage would, if the action went far enough, be nearly normal to the rigid planes in both cases. At least, so it seems to me, at present. Now this may be what the other party has in mind.

I wish you would privately send me references to the strongest papers by *others* than yourself on such application of analytical mechanics to geology. For I want to say to the director how he ought to rate you as an authority upon such questions.

I want the money he offers me for my opinion BAD, but I fear I can't take it in that form. If he were to offer me so much per annum to answer all such questions, say $600 or $900, *that* I could take without hesitation. But it would be quite absurd when you are there who are much more competent.

I owe you $50. I have been instrumental in placing the bonds of a company which is to create a new water power in St. Lawrence

County New York, leading the water of the southern channel of the St. Lawrence south of an island whose name escapes me into the grass river and thus getting an available fall of 40 feet and developing 100,000 horse power. I am promised $30000 of the stock for what I have done; and am sure it will pay very large dividends beginning in say, 18 months from now. When I get that, if I do get it, as I fully believe I shall get about that, I shall be able to pay all debts. I have also a place in [Bucks] County worth $15000 or $20000 on which there is something less than $2500 in mortgage and lien and I am trying to raise a loan on that to tide me over. But owing to a conspiracy to ruin me and get the property away from me which exists up there and the fact that most of my socalled friends are delighted to see me in trouble and will put themselves to some little inconvenience to increase it, I have the greatest difficulty in getting any money or employment. I have a few very good pictures too; but this is no time to sell pictures. If I could raise $1000 for a year, the best brokers assure me I could by putting half that on improvements etc. certainly add $5000 to my mortgage there.

I mention all this because of my owing you that sum and I want to explain why I have not paid and that I expect to pay. But I am in great danger of losing my place in the country if there were any department in Washington which wanted such services as I could render, I should be glad to take a very small salary, especially if the services were such that they would leave me time for some outside work.

Another reason why I mention the matter is that when our company gets the water power we intend to use a great part of it in making carbide of calcium. And I want to know whether you can direct me to information about the limestones of St. Lawrence county. They want to be free from silica. I have an idea that magnesium limestone might prove the best. It is true that owing to the fusability of MgO as compared with CaO a higher temperature is required; so that Moissan says that MgC_2 cannot be made in the "Four électrique." But I think it could be an ingredient of a carbide from magnesium limestone, and its lower atomic weight would be an important consideration, not only because the carbide will be carried all over the world, but also because, if we took a patent on the process, the prestige of making richer carbide than anybody else would be valuable. For the same reason I would avoid strontium notwithstanding its easier fusability. There is so much limestone up there, that I think we can't fail to find just what we want without going far.

I have some patents connected with the use of acetylene which will be valuable when it comes into use as it unquestionably will in a year or two. To find uses for electricity is now a great field for invention.

yours faithfully
C. S. Peirce

Merrill Collection, Library of Congress

C S. Peirce to William James

84 Broad St. 1897 March 13

My dear William:

Your letter and the dedication and the book[7] gave me more delight than you would be apt to believe. The note came day before yesterday. I got the book last night. I have read the first essay which is of great value, and I don't see that it is so very "elementary" as you say, unless you mean that it is very easy to read and comprehend, and it is a masterpiece in that respect.

That everything is to be tested by its practical results was the great text of my early papers; so, as far as I get your general aim in so much of the book as I have looked at, I am quite with you in the main. In my later papers, I have seen more thoroughly than I used to do that it is not mere action as brute exercise of strength that is the purpose of all, but say generalization, such action as tends toward regularization, and the actualization of the thought which without action remains unthought.

I have learned a great deal about philosophy in the last few years, because they have been very miserable and unsuccessful years, terrible beyond anything that the man of ordinary experience can possibly understand or conceive. Thus, I have had a great deal of idleness and time that could not be employed in the duties of ordinary life, deprived of books, of laboratory, everything; and so there was nothing to prevent my elaborating my thoughts, and I have done a great deal of work which has cleared up and arranged my thoughts. Besides this, a new world of which I knew nothing, and of which I cannot find that anybody who has written has really known much, has been disclosed to me, the world of misery. It is

[7] *The Will to Believe and Other Essays in Popular Philosophy.*

absurd to say that [Victor] Hugo, who has written the least foolishly about it, really knew anything of it. How many days did Hugo ever go at a time without a morsel of food or any idea where food was coming from, my case at this moment for very near three days, and yet that is the most insignificant of the experiences which go to make up misery? Much have I learned of life and of the world, throwing strong lights upon philosophy in these years. Undoubtedly its tendency is to make one value the spiritual more, but not an abstract spirituality. It makes one dizzy and seasick to think of those worthy people who try to do something for "The Poor," or still more blindly "the deserving poor." On the other hand, it increases the sense of awe with which one regards Gautama Booda. This is not so aside from the subject of your book as it might seem at first blush, because it implies that much has led me to rate higher than ever the individual deed as the only real meaning there is the Concept, and yet at the same time to see more deeply than ever that it is not the mere arbitrary force in the deed but the life it gives to the idea that is valuable.

As to "belief" and "making up ones mind," if they mean anything more than this, that we have a plan of procedure, and that according to that plan we will try a given description of behavior, I am inclined to think they do more harm than good. "Faith" in the sense that one will adhere consistently to a given line of conduct, is highly necessary in affairs. But if it means you are not going to be alert for indications that the moment has come to change your tactics, I think it ruinous in practice. If an opportunity occurs to do business with a man, and the success of it depends on his integrity, then if I decide to go into the transaction, I must go on the hypothesis he is an honest man, and there is no sense at all in halting between two lines of conduct. But that wont prevent my collecting further evidence with haste and energy, because it may show me it is time to change my plan. That is the sort of "faith" that seems useful. The hypothesis to be taken up is not necessarily a probable one. The cuneiform inscriptions could never have been. . . .[8]

William James Papers, Harvard University

[8] Letter incomplete.

COPE AND MARSH: THE BATTLE OF THE BONES

❁

Paleontology was one science where nineteenth-century America could clearly match Europe. Fossil remains had been found and studied in America since the previous century. With the efflorescence of geology and natural history, nineteenth-century Americans made many first-rate contributions to the study of extinct forms of life. Paleontology was a new and major science then, a source of intellectual excitement among the educated public. With the coming of Darwinian evolution, the science acquired an even greater significance; here were possibilities of proving or disproving evolution and natural selection, and men all over the world scanned the rocks for remains of missing links. In America the last century produced three great paleontologists—Joseph Leidy (1823–91), Edward Drinker Cope (1840–97), and Othniel Charles Marsh (1831–99). Cope and Marsh became embroiled in a feud of titanic proportions, partly a result of the personalities of the two combatants, partly a consequence of the growing pains of their discipline. Eventually, the Congress, the National Academy of Sciences, the U.S. Geological Survey, and a large portion of the nation's geologists and paleontologists were brought into the fray.

At first paleontology was a rather disorganized intellectual venture. Bones were found, sometimes by chance travelers, and described by naturalists. As the surveys and expeditions, public and private, became more common, a paleontologist might come along to dig in promising locations. Usually a field man sent the specimens home to a specialist in one of the scientific centers back East, as Ferdinand V. Hayden did to Leidy. If the paleontologist wanted to do more than uncover an occasional fossil specimen and to describe systematically the past fauna of a particular region or a particular kind of extinct life such as the dinosaurs, it was increasingly necessary to establish a fixed relationship with the parties exploring promising regions, usually with a state or federal geological survey. Where the paleontologist was once an amateur naturalist, in the post-Civil War years he was likely to be a highly specialized member of an organized natural history project.

Edward Drinker Cope was a Quaker of a decidedly unpacific temperament. His father was wealthy and provided his son with a good, if somewhat unconventional education. Endowed with a brilliant, wide-ranging mind, Cope made valuable contributions to practically every branch of natural history, but is best known for his work in vertebrate paleontology. Unlike Leidy and Marsh, he was a theorist, being the leading American Lamarckian of his day. Covering so much ground and working with great speed, Cope sometimes committed errors and blunders in his haste. Once Leidy, his fellow Philadelphian and former teacher, discovered that Cope had reconstructed a skeleton with the skull at the wrong end of the vertebral column. Marsh, his great rival, never let Cope forget that incident. Cope started out based in the American Philosophical Society and the Academy of Natural Sciences of Philadelphia; he ended up widely disliked in his own organizations and with few friends among his contemporaries. But young men, like the Princetonians Henry Fairfield Osborn (1857–1935) and William Berryman Scott (1858–1947) and his students at the University of Pennsylvania during the last years of his life, took to him. Although he lost his battles with Marsh, through them he won the war, for he is widely recognized as the most brilliant of the three. And many of his blows against Marsh were fatal to that scientist's reputation, at least as a human being.

Othniel Charles Marsh was the nephew of George Peabody (1795–1869), the financier and philanthropist, and was, like Cope, independently wealthy. In contrast to the wide-ranging Philadelphian, Marsh was highly specialized and worked with great care. He had the knack of spotting the main chance and seizing it. Marsh made two very important contributions to the developing body of paleontological evidence for evolution: the discovery of extinct birds with teeth, a kind of missing link, and the tracing of the development of the modern horse from his remote ancestor, Eohippus. Marsh was highly respected by his contemporaries, or at least he was a successful politician and organization man, serving as President of the National Academy of Sciences from 1883–95. But Marsh had one fatal flaw—his relations with his subordinates at the Peabody Museum at Yale. Marsh did not teach (until 1896 when he was financially embarrassed, Marsh did not receive a salary) and had no students. The assistants at the museum, many of them good scientists in their own right, were regarded merely as hired hands who worked at small bits of his research so that they could not either rival Marsh or leak information to Cope. The museum was a research factory for the glory of Marsh, not an institution of colleagues engaged in research. Leaving no followers, Marsh's reputation went into an eclipse from which he has only partially emerged.

Both men were rather nasty or at least difficult to deal with in many ways. In the course of time as the feud progressed, they became increasingly suspicious, secretive, and paranoid. They had the misfortune to work in the same field in the same country at a time when priority conflicts in science were especially rampant because of a lack of accepted institutions and protocols for resolving these disputes. Perhaps the most severe struggles of this nature occurred in taxonomic fields where each description could yield a kind of immortality to the discoverer when his name was attached to the formal appellation.

In these fields the conflicts were further aggravated by the importance attached to physical objects—that which was being described. The rivalries soon centered on who would obtain possession of the bones, plants, skins and rocks. And the struggle to acquire these became more than the rivalries of individual collectors but full-scale battles between museums. In his lifetime Louis Agassiz outdistanced his rivals in America in sheer acquisitiveness for his Museum of Comparative Zoology at Harvard. (At the same time Spencer F. Baird of the Smithsonian, who had access to the collections brought back by the government parties, was quietly laying the foundations for the U.S. National Museum.) Before he died Agassiz started negotiations with James Hall of Albany for his superb collections, particularly rich in invertebrate paleontological specimens. His son, the zoologist Alexander Agassiz (1835–1910), could not complete the deal. Hall sold out for $65,000 in 1875 to Albert S. Bickmore (1839–1914), a rebellious student of the older Agassiz who founded the American Museum of Natural History in New York.

But Cope and Marsh, not Bickmore, were to dominate the scene for the next two decades in paleontology. They were robber barons trying to corner the old-bones market. Cope was at a disadvantage since the old Philadelphia scientific societies were imperfect vehicles for his ambitions. The first stage in their rivalry was the elimination of Joseph Leidy, a small operator trying to compete with two great corporations. Leidy lacked the financial resources, and he eventually lost his connection with the Hayden Survey to Cope. Leidy tried to continue, going out into the field and imitating Cope and Marsh in the practice of telegraphing back discoveries to gain priority. As the struggle between the two feudists became more bitter, Leidy withdrew from a field he described as no longer fit for a gentleman.

Cope and Marsh were not above suborning each other's associates nor of outbidding and defaming each other. Nor were they scrupulous about the rights of others. A story is told of old Professor Guyot of Princeton arranging with a farmer to buy a specimen discovered on his land. Hearing of this, Marsh chartered a train and rushed down to New Jersey, bought the fossil at a higher price from the discoverer who was waiting at the railroad station for Guyot. When Guyot arrived, the station was bare.

Marsh had an advantage over Cope because *Silliman's Journal* was published at New Haven. James Dwight Dana, while printing Marsh's pieces, soon detested him for involving the periodical in the unseemly squabble. Cope countered by purchasing the *American Naturalist*. Marsh's most serious blow at Cope came when the existing geological surveys were combined into the U. S. Geological Survey in 1879. He was appointed paleontologist of the new organization, thus cutting off Cope from his principal source of materials at a time when unwise investments had taken away his fortune. Cope also felt that Marsh and the Survey's director, John Wesley Powell, were conspiring to prevent the publication of his work for the Hayden Survey.

From 1879 on Cope was engaged in almost open combat with Marsh, the Geological Survey, and the National Academy. Cope lined up with those who were against Powell and his programs, quietly feeding information to hostile Congressmen and newspapers. He probably expected that Marsh would have his downfall in 1885–86 when the Allison Commission, a joint body of the House and Senate, investigated the scientific bureaus of the federal government. From the evidence of such letters as that to Osborn of October 27, 1885, which appears below, Cope expected a revolt of Marsh's assistants to aid in the destruction of his rival. In that year Scott went to New Haven to have a secret meeting with the dissidents at the Peabody Museum under the ruse of attending a Yale-Princeton football game. Marsh and Powell survived.

In 1889, whether deliberately or not, the Survey struck back. Cope was requested to return the fossils gathered during his association with the Hayden Survey. Cope had not received a salary; he assumed that the fossils were his private property. Particularly enraging to Cope was his belief that Marsh at the same time was profiting intellectually from access to fossils that were public property and was enriching his own personal collections at Yale. In 1890 Cope arranged for the airing of his charges against Marsh in the newspapers, especially the claim that the assistants, not Marsh, were the authors of monographs for which he took credit. But Marsh was not unprepared; he replied with a carefully gathered catalog of Cope's sins. The enemies of science rejoiced; Marsh was eventually forced out of the Survey; American scientists were horrified at the undignified squabble, especially since there was a high degree of truth in the charges of both parties. Cope, in financial distress, sold part of his collection to Osborn (now at the American Museum of Natural History) for $32,000 in 1895. After his death the remainder was purchased by the Museum for $28,550. Osborn was now launched in a notable career as the leading museum man of his generation.

Leidy, Cope, and Marsh were great contributors to paleontology. Prior to their work only 98 genera and species of North American fossil vetebrates were known; to this number they added 2193 genera and species. Of this total Leidy contributed 375, Marsh 536, and Cope 1282.[1] Leidy also did significant work with invertebrates and deserves great credit for many priorities in vertebrate paleontology achieved under very difficult circumstances. Marsh was notable for the quality of his findings. Leidy refused to generalize—both from personal inclination and the incomplete state of his fossil evidence. Cope had few qualms about theoretical speculations, often with little evidence and usually tinged with a desire to introduce purpose into evolution by way of Lamarck's inheritance of

[1] From H. F. Osborn, *Cope: Master Naturalist,* p. 20.

acquired characteristics. Marsh, the specialist, did not theorize. He had a generalization, Darwinian evolution, which he accepted and which gave direction to his work.

Edward Drinker Cope to Othniel Charles Marsh

Haddonfield, N.J. 5/4 1868

My dear Prof. Marsh

The express agent promises to come for thy boxes immediately, on my requesting it, at thy departure from this place. This he neglected to do, but comes, I hope certainly, tomorrow morning. The delay is to be regretted, but I have in the meantime found what is perhaps a branched sponge which I desire for the Academy, but lend to thee if it be of any interest to thee. It is from Barnesboro. I put it in the larger box, just underneath the figure of a wine glass.

I have found the tibia that accompanies the humerus of strange form and an other tibia like the one I had with the latter when thee was here. There are obviously two species, which I call *Euryomus rectimanus* and *Euryomus grallavius*. The large vertebra is positively reptilian and from the Cretaceous. It is a Cimoliasaurus of a new and large species which I call *Cimolius aunis maximus*.

Gabb tells me the Frizonia is common in the cretaceous of N. J. Has thee come across the little spiral shell. It was in a flat box on some blue cotton sheeting.

We are all well here, and still hoping for *good weather*. With Kind regards I am thy friend

Edw. D. Cope

Cope to Marsh

Haddonfield 2/28 1870

My dear Prof. Marsh,

Please have my thanks for the proofs of the bird M.S. I am much interested in them and think thee remarkably fortunate in getting the specimens.

As we sometimes take the liberty of alluding to each others "great mistakes," I would make one criticism on the diagnoses of the birds, i.e. that thee does not "come up" to the generic relationships nor show any reasons why the genera named, are different from those existing. This is the Leidyan method, and one that rendered one *written science* a riddle, and leads foreigners to underrate our abilities (?).[2] Characters or nothing!

I lately described a new Mosasaurus at the American Philosophical Society, under the name of *Mosasauris carrhrus*, it of the Second grade of size between M. maximus and M. fulciatus. Allied to M. depressus.

I have met with still more delays in printing my book, and thee would scarcely believe that the chairman of the publication committee is trying to prevent my having the plate the society granted, and two others granted before! Hence the delay. I will however send some sheets erelong.

I have just made a trip through the marl country N. and S. of here and obtained some bones.

With Kind regards
E. D. Cope

Cope to Marsh

Philadelphia 5/31 1871

My dear Prof. Marsh,

I have just r[eceived] the American Journal of Science Arts and read your article on Cretaceous and Tertiary Reptiles with great interest. Your discovery of [posterior] extremities in the *Pythonomorpha* is an important point gained. Your new species of the same group please me much, and I am glad how to see you define the genus *Edestosaurus* so clearly. This thing of *defining* of genera is you know, a point I take interest in, and our science will grow or stand still just in proportion as this is done or not done. I judge by your text that you have *Clidastes* also.

By the way I see your printer has also made a discovery i.e. that Mosasaurs and Crocodiles are "Serpents"!

With Kind regards I am etc.
Edw. D. Cope

Cope to Marsh

Haddonfield 1/20 1873

Dear Prof. Marsh

I send you some small specimens I recently received from Kansas as having been abstracted from one of your boxes! Of course they are yours.

[2] Cope's question mark.

Your bird with teeth is simply delightful. *Vae evolutionis opponentibus!* De mortuis nil nisi boneum! . . . [Woe to the opponents of evolution! Speak nothing but bones of the dead! . . .]

Marsh to Cope

Yale College New Haven Jan. 27th 1873

Dear Prof. Cope.

I am glad you fully appreciated my bird with teeth and I hope soon to send you some photographs of it.

Your paper on the "Proboscidians" came 20th ultimate with postmark 18th Jan, although bearing date 16th. Why don't you send your papers more promptly, as I invariably do. I am willing to accept as publication even an uncorrected proof, (as we agreed) when received, so far as I am concerned, and Leidy promised to do the same.

I regard it as more really publication to send on day of issue to you and Leidy than to 50 others who are not interested in the subject, although the former would only by courtesy be regarded as publication. Dana, Verrill,[3] and several others with whom I have spoken, are unanimous in saying that extras sent out in advance of the Periodical can only be considered when they are made acceptable to those working in the same department.

I notice in your paper that you give the date Aug. 24th 1872 to the genus Tinoceras. The name, however, was in the erratum of my pamphlet on Mammals which I mailed to you Aug. 19th and to many others. It was in print two or three days before. The Kansas fossils you sent came all right. Where are the rest? and how about those from Wyoming?

The information I received on this subject made me very angry, and had it come at the time I was so mad with you for getting away Smith,[4] (to whom I had given valuable notes about localities etc.). I should have "gone for you," not with pistols or fists, but in print. I came very near publishing this with some of your other transgressions including a certificate from Mr. [O. B.] Kinne[5] but my better judgment prevailed. I was never so angry in my life.

Now don't you get angry about all this but pitch into me with equal frankness if I have done anything you don't like.

[3] Addison E. Verrill (1839–1926), a zoologist at Yale.
[4] Sam Smith, a former field man of Marsh's.
[5] An eastern railroad official who gathered fossils for Marsh and resisted Cope's lures.

In haste yours very truly

OC Marsh

Cope to Marsh

Philadelphia 1/30 1873

My dear Prof. Marsh

I wish you had mentioned to me about missing specimens from Kansas, Wyoming etc. When the first suspicion crossed your mind that I knew anything about them. It is far more irritating to me to be charged with dishonorable acts than to lose material, species etc.

I never knew of any losses sustained by you, or specimens taken by any one, till those were sent me that you now have. Should any such come to my hands I will return them as I did the last.

As to Mr. Kinne, he has, as I have long suspected, misrepresented me. He tells of fish teeth etc. which he lent me which were not returned. If my [publications] and cabinet are worth anything his statement is false.

On the other hand some appropriative person has stolen *Chlorastrolites* Hyposaurus jaw etc. from me.

All the specimens you obtained during August 1872 you owe to me. Had I chosen they would all have been mine. I allowed your men Chew[6] and Smith to accompany me and at last when they turned back discouraged, I discovered a new basin of fossils, showed it to them and allowed them to camp and collect with me for a considerable time. By this I lost several fine things, although Smith owed me several days work. My two weeks spent with you in Jersey and exhibitions of *unpublished* material are set off by your refusal to let Mudge[7] go with you in Kansas in 1871 and your charges of dishonorable conduct etc.

Now as to a man of honor I request of you

1st. To correct all statements and innuendoes you have made to others here and elsewhere, as to my ?[8] dishonorable conduct.

2d. To inform me at once if others make such charges to you, about me.

Hoping you will find this much on the credit side of your account I am yours for Worth

E. D. Cope

[6] John W. Chew, another Marsh field man.

[7] Benjamin F. Mudge, then at Kansas Agricultural College, was the one who provided Marsh with the bones of the "birds with teeth." Cope's point is obscure; Mudge continued to collect for Marsh after 1871.

[8] Cope's question mark.

Marsh to Cope

[no date]

Dear Prof. Cope

In reply to your letter of the 30th ultimate I have only to say that

1st. I desire most sincerely to be on friendly terms with you

2nd. If I have, by word or deed, done you the slightest injustice, or should in future unintentionally do so, I will promptly make due amends as soon as I know of it.

If you can truly say the same, we have at once a basis of agreement that will prevent any serious misunderstanding in future.

As to the past, I will say frankly that I feel I have been deeply wronged by you in numerous instances. These wrongs I have usually borne in silence, and have even defended you when others have spoken against you. After the Smith affair last summer I made up my mind that forbearance was no longer a virtue.

In regard to Smith, let me remind you that I had spent no little time in teaching him to collect fossils; had entrusted to him most valuable information about localities, (including those East of the Green River); had given him an outfit, and engaged him for the season at his own price. When in June he had a good offer to go on a Military Expedition, he declined, saying that he could not go without my consent. You, however, enticed him away, even before he had shipped his specimens as directed, and they were then delayed with great loss to me. I would not have done this to you for all the fossils in Wyoming.

This act of yours created a strong prejudice against you at Fort Bridger, among both officers and civilians. Three separate parties promptly informed me of it, and denounced you for it in strong terms. I was likewise informed of your efforts to examine the specimens collected for me, and it was believed then that you obtained some of them. Now for all this you have only yourself to blame. I had nothing whatever to do with it.

In regard to the Kansas fossils, let me say with equal frankness that I had lost some valuable specimens, and had some reason to think they were in your possession. Could I have done less than to give you a chance to explain the matter? You have said distinctly that you have neither Wyoming or Kansas fossils of mine, and I have, therefore, nothing more to say.

One part of your letter I do not comprehend, viz. the reference to Chlorastrolites etc. If you mean Mr. Kinne I will simply say I do not believe it. If you refer to me, the charge is utterly false and you must know it.

[no sig.]

Marsh Papers, Yale University

E. D. Cope to Alfred Cope

Washington D.C. 3/16 1873

My dear Father

I had intended writing to Fairfield before coming here, but I finished my home work barely in time to take up that awaiting me here. I am now nearly through with this, and hope to be at home by third day evening. I would have been glad to go to Fairfield on th[e] evening that I lectured in Germantown but cousin J. S. Haines had invited me so often, and I have so frequently been unable to get there, that I thought better to go with him, especially as I had to leave early the next morning. I find abundant facilities here, and friendliness on the part of the officers. Some one has just endowed a chair of Natural History at Princeton College to be called the Henry Chair, and Prof. H. recommended me to Prof. McCosh[9] to fill it. The latter objected to my evolution sentiments, for those views are much condemned at Princeton. I have not much intention of fixing myself there, as the hours and work generally will probably require too much time; but I may find the University of Penna. better, especially as it is nearer home. One or the other I will probably undertake.

Friend's meeting here is situated at some distance from the centre of the City, and is held in a room attached to the colored mission or school. It is quite small, there being not more than ten regular attenders. Strangers frequently drop in. Yardley Warner passed through yesterday.

The corps of workers here varied from year to year, but a few remaining permanently. There are two here who commenced with me; several have come in since, and the style and standing of the corps has improved. Some are doubters or unbelievers, but only one at all frivolous. There are two consistent Methodists who do their faith credit, and one at least of them will take a high place in science if he perseveres. It seems very hard for some to accept any

[9] James McCosh, President of Princeton.

evidence except that of the senses, and having no other, they deny supernaturalism or class it with delusions. It is hard to reason with such men, especially if they be tolerably moral, though they very seldom are even approximately so. One can only assert, and relate experiences, which is generally "casting pearls before swine." This I learned some time ago, and am accordingly careful.

I suppose my paper on the fossil Proboscidians of Wyoming is now out. In it Prof. Marsh will find something to digest, though nothing indigestible. I hope to make matters so plain that he will have to swallow them whether he will or no. His charges are quite strong and I may have to prove them false by a special note, but I hope the internal evidence of the descriptions, plates etc. will be sufficient.

I hear that Mother does not improve. Remember me affectionately to her. I hear all well from Annie at Haddonfield.

With love thy son
Edward D. Cope

E. D. Cope to A. Cope

Fort Bridger Wyo. 10/3 1873

Dear Father

I have just received thy note containing $200.00 for which many thanks. I have been looking after my livestock and waggon here, as well as the fossils. I have collected by proxy. I have sold my whole "outfit" for some months work in collecting fossils, which will probably do very well, amounting to as many fossils as I could get by spending a very good price received for the stock etc. The material lying here for me contains some good things, among others another cranium of *Soxolophodon cornutus*. I found on examining some books here, that *Mastodon* I procured in Colorado, is a new species somewhat like the common one, but peculiar and from an older formation. I called it *M. proavus*.

Tomorrow I go to Evanston. At this place there is a great puzzle as to the limits of the Cretaceous and Tertiary formations. Indeed this whole country seems to furnish closer connections between the two than any other; hence there are differences of opinion as to the boundary line. Last summer I made some important progress in proving that what many called Tertiary was really cretaceous. This dissatisfied some who thought their evidence more weighty. Marsh was wroth because I did not credit him with the discovery (he had seen something of the kind at a locality far off and disconnected) while another man who had decided both ways (he didn't know which) claimed the discovery with vigor! So if you do a good thing it is either not true or done long before! However, I have more evidence still which will no doubt disgust somebody and make some others ashamed of their barking. I find Marsh at a discount out here where his manners have produced the same impression that they have East. Progress is only made in spite of annoyances from parasites and jackals, but it is a good discipline to learn to bear with them, as our great example did with the Jews. However he denounced them sometimes. The paper is out. Love to all. Thy affectionate son Edw. D. Cope

Cope Papers, American Museum of Natural History

Cope to H. F. Osborn

Philadelphia 10/27, 1885

Dear Prof. Osborn:

The four men who have left Marsh wish to place themselves and him right before the Scientific public. They have found M. to be more of a pretender than even I had supposed him to be. I have recently seen a statement in M.S. which sets forth a number of things which are worse than I had supposed. It is now clear to me that Marsh is simply a scientifico-political adventurer who has succeeded, in ways other than those proceeding from scientific merit, in placing himself in the leading scientific position in the country. It is now perfectly certain that he, M., has not written either of the quarto books that bear his name, and it is doubtful whether he has written much or any of his 800 papers.

I consider that this career is a disgrace to our scientific community, and one that we ought to wipe out. The American National Academy of Sciences "will stink" in the noses of corresponding bodies in other countries.

This winter there is to be a thorough ventilation of the Geological Survey at Washington. Marsh will probably try and get the National Academy to boost the organisation. There is a good deal of Marshism in other departments than the Vertebrate Pal., as I am told. In any case such an opportunity of placing Marsh in his true position will not soon occur again.

The 4 men at Yale are anxious to publish in both Europe and America their statement, and I naturally am willing that they should. There is however one difficulty. Dr. [George] Baur, who is as important as any of the four, and furnishes a good deal of the backbone, is in Marsh's debt some $650, an[d] he can do nothing till that is paid as he is a married man. I want to find some one to lend him the money or part of it, because if it is paid out of his wages, he cannot pay it till next April, which will be too late for the paper to be of any service in the investigation. Although I am entirely impecunious, I have taken upon me to try and raise $200 of this amount. I do not yet know whether I can succeed in getting the amount or not. Supposing that Baur shall be free by Dec. 1st there remains $400 to raise. Now Dr. Baur will probably ask you whether you can lend him this amount, or part of it? Certainly this winter and *early,* is the time to straighten up this matter and put our vertebrate paleontology on a decent basis. The publication of the paper will be a blessing to American Science generally, for the demoralizing effect of Marsh's success is incalculable. Besides he has plenty of means to work with if he chooses. He is not contented with this however. He has completely suppressed my work. Not a thing has been done towards final publication for now *two years,* and hope of anything being done is *growing* steadily less. This will prove not only an inconvenience to me. Paleontologists generally will want plates of my species which have now remained, some of them without illust[r]ation for several years.

Please give my best regards to Prof. Zittel. Who will do his vertebrate Paleontology now Kavelevsky is dead? [10]

With best regards I am yours

E. D. Cope

Marsh to Osborn

New Haven, Conn. February 27, 1890.

Dear Professor Osborn,

I have received to-day the memoir on the Mammalia of the Uinta Formation, by Scott and yourself, and write to thank you both for sending it to me.

I have as yet only had time to glance it over, but hope soon to read it carefully, as the subject interests me very much.

You know of course that I discovered the Uinta Basin in September, 1870, and made the first collections there under great hard-

[10] Karl A. von Zittel (1839–1906), a leading German paleontologist and geologist. W. O. Kowalevsky (1843–83), a Russian paleontologist, was, like Cope, a Lamarckian. Cope had charged Marsh with stealing from Kowalevsky's work on fossil horses.

ship and danger. I first gave the name "Uinta Basin," first determined its Eocene age, and its position at the top of the series. I have a very large collection of its fossils, in addition to those I have described, and will perhaps write you later in regard to some points in which we do not seem to agree, especially in regard to nomenclature.

Let me now, however, ask one question about the date of your memoir, which I do not understand. It bears the date, "August 20, 1889," after the words, "reprinted from the Trans. Amer. Philos. Soc., N.S., Vol. XVI, Part III." Does this mean that your memoir sent to me was printed at that time, or that the transactions themselves were printed at that date, and your separate memoir reprinted later? The transactions named have not been delivered to the members here, or to our library.

I have noted the date of receipt on your memoir which came to-day, as I do in all such cases.

Yours; very truly,
O. C. Marsh.

Osborn to Marsh

July 11th [1892]

Professor O. C. Marsh.
Dear Sir:

Mr. Jesup[11] has requested me to answer your letter to him of June 30th. Please refer to my letter *to you of March 9th, to which I have received no answer.*

The course of the Museum in regard to collecting is governed by Mr. Jesup's official letters to you. All I find take the same ground which is most distinctly stated in that of February 19th, after consultation with the Committee of the Trustees: "We agree with him (i.e. Prof. Osborn) that we cannot promise to confine our explorations to any particular localities. We want to make the collections as representative as possible; therein rests their value to Science."

Also by my letter of March 9th "I have given explicit instructions to my collectors to respect the rights of other parties and carefully avoid difficulties in the Field."

These letters were written after careful consideration.

In my informal note to Mr. Jesup of March 7th, which he inclosed

[11] Morris K. Jesup, President of the American Museum of Natural History.

to you, I wrote of the plans which were formed at the time. The contents of that note contained no promise or agreement with you. As I state above we have made no agreements whatever as to localities.

None of the trouble you have anticipated in the field has arisen or will arise. Dr. Wortman has seen Mr. Hatcher[12] and come to an understanding with him, they will respect each others rights.

We have received a small collection of Mammalian teeth which will be made accessible to you for examination and study whenever you wish to see it. I do not know the exact localities in which they were found and have written to Dr. Wortman to inquire. Even if they were found in the same localities as those in which your collections were found, they remain the property of the Museum. Our Puerco Collection has just been received; it was made by a man formerly in Professor Cope's employ and in the exact localities discovered by him. It would be as reasonable for Professor Cope to claim our Puerco Collection as it is for you to claim the Laramie Collection. And I am surprised that you advance such a claim. I am

Truly yours,
Henry F. Osborn

Osborn Papers, American Museum of Natural History

[12] Jacob L. Wortman, a paleontologist, was collecting for the American Museum. John Bell Hatcher was Marsh's most renowned digger; in 1893 when the Geological Survey funds were cut off, he left Marsh's employ.

THE RISE OF PHYSICS

Few Americans in the year 1900 would have predicted that physics would become a major scientific activity in twentieth-century America. There simply was no evidence of any ground swell in that discipline. Even as late as 1940 many Europeans would have scoffed at the assertion that America would supplant the traditional centers of leadership in the field of physics. Although America had produced physicists of note, they were relatively rare and the national style did not appear to encourage the development of theoretical fields. In retrospect it is clear that a community of research physicists was developing in the United States since at least the latter decades of the previous century.

As geography ceased to provide a framework for the investigations of natural historians and geophysicists, the particular pattern of research prevalent in America up to roughly 1870 became increasingly anachronistic. Because of its prestige, Americans gave fairly generously of their national wealth to the support of astronomy. By the end of the century it was probably the field in which Americans rated best in comparison to the Europeans. After a promising start meteorology became bogged down by too early a dedication to applications with too little basic research. Geology continued as a great, active scientific field in America. Pure mathematics was badly neglected. Chemistry, comparatively neglected during the heyday of the geographical sciences, did expand, first under the stimulation of the needs of agriculture and medicine, later from the impetus of industry. In general, chemistry in America in the latter decades of the nineteenth century was heavily oriented towards applied research. Achievements in America generally did not compare favorably with the leading work in Europe.

In contrast the natural sciences flourished. Widespread amateur interest continued and assured professional biologists of a favorably inclined audience for requests for support. By the end of the century the growth of medical and agricultural research provided additional funds, laboratory facilities, and chances of employment. As a result there developed a rather large, if amorphous, American biological scientific community which impressed European observers after the turn of the century.

By 1920 approximately 20 per cent of the articles written in the world on physiology were by Americans.

The development of physics was quite different. Instead of fairly wide support for a large number of investigators, a few outstanding physicists appeared, won modest backing for their researches, and stimulated young men to enter the field. Their contemporaries of lesser stature were encouraged by the examples of compatriots who were peers of the great names in Europe. In retrospect, the great names in the physical sciences in late nineteenth-century America—Rowland, Michelson, Gibbs—far overshadow their contemporaries in the biological sciences in America.

They were not, of course, the first physicists produced in America. Nor did they appear miraculously without any antecedents in the American scientific community. They had been preceded, for example, by physicians who tried their hands at electricity or some other fashionable physical topic. Of this group the two most important as harbingers of later developments in the physical sciences are John William Draper (1811–82) and his son Henry (1837–82). They are discussed in the next section. Others were engineers by training with a taste for pure science, like Henry Augustus Rowland. Still others, like Joseph Henry and A. A. Michelson, had roots in the geophysical tradition. Only Gibbs defies successful placement in some antecedent scientific trend.

The Drapers—The Transition to Professionalism

Both Drapers taught at what is now the Medical School of New York University. Although they made contributions to subjects within the bounds of the medical curriculum, their greatest achievements were far removed from the theory and practice of their professions. John William Draper, essentially a chemist, early developed an interest in photochemistry. After learning of Daguerre's work, Draper was the first to take a photograph of the human face. Later he was the first to obtain a photograph of the moon. Given his great interest in radiant energy and in the application of photography to astronomy, Draper not surprisingly turned his attention to the analysis of the sun's radiation, producing the first photograph of the solar spectrum. The elder Draper had one additional characteristic which brought greater contemporary fame than all of his scientific discoveries. Filled with an almost religious belief in science and in the idea of progress, Draper wrote highly successful histories in support of those views. His final literary production was an account of the antagonism between religion and science in which he read back into the past the intellectual battles of the mid-nineteenth century.

Henry Draper's career paralleled his father's and carried the analysis of light significantly forward. John William Draper was apparently one of the first Americans to work with diffraction gratings to produce a spectrum. His son Henry, a superb designer and constructor of telescopes, continued the research on the solar spectrum, eventually demonstrating the presence of oxygen. But his greatest achievements were in stellar spectroscopy where he succeeded in making the first photograph of a star's spectrum showing Fraunhofer lines. After Henry Draper's death his wife endowed the preparation of the monumental Henry Draper catalog of stellar spectra by Harvard Observatory.

Both Drapers were highly regarded in their day by leading scientific figures. Although their accomplishments were substantial, they are now often condescendingly characterized as amateurs. In a sense they were. Neither went to a graduate school, neither belonged to any guild of physicists, neither taught the subject. They used their own funds and did not seek outside support. Judged in terms of skill and devotion to their work, judged in terms of their contributions, they were clearly more than mere amateurs. In comparison to many of their contemporaries who grubbed conscientiously in the classical fields, it was the Drapers —physicians gone astray—who were in the mainstream of research in physics.

Why are the Drapers so little known? In part it is because later scientists did not recognize them as part of the guild. Another reason is that experimental work, unless producing the sensational, is likely to loom less in retrospect than theoretical work, which generalizes the results of many experiments. Medical school did not give the Drapers the mathematical background necessary for meaningful theoretical work in physics by the late nineteenth century. But most damaging to their future reputation was the near impossibility of convincing any large number of Americans of that day of the importance of photographing the spectra of the sun and the stars.

John William Draper to Henry Draper

Wednesday evening, Nov. 23rd. 1870

My Dear Henry

Yesterday morning I set out to see Mr. de La Rue[1] but it rained so hard that I was driven back and got pretty wet in the attempt. In the afternoon it cleared off and so I sallied out again. Fortunately I found him in. He recognized me at once, and gave me a very cordial welcome telling me that he should pay his respects to Aunty and Antonia.

He gave me an invitation to the Annual dinner of the Astronomical Society on Dec. 9th and a letter to Dr. Huggins.[2] He

[1] Warren de La Rue (1815–89), an English astronomer.

[2] Sir William Huggins (1824–1910), the English astronomer, perhaps the most noted contemporary of the Drapers in the use of spectroscopy and photography in astronomy.

asked very kindly after you and inquired as to your doings. So I gave him a description of the new telescope. He said he had made some trials as the relative photographic goodness of Speculum metal and Silvered glass and had been brought to the conclusion that the former is the best. He believes that many of the active rays get through the silver film and it appears to me that this may be so, as it transmits plenty of blue light.

I said to him, now Mr. de La Rue, who examined the Australian telescope told me candidly that it really was for we have heard in America that the Australians were disappointed with it. Well, he said, I cannot give you a better proof of what it can do than that I saw Y. Andromede not only double but triple through it. It gives a star not as a disc with diffraction fringes as a refraction would do but as a bright speck with minute irradiations, they are however very minute. Its defraction is superb. The reason of the difficulty in Australia was this. Grubb covered the mirror with a varnish of shellac to prevent tarnishing on the voyage and when they attempted to remove the varnish instead of using concentrated alcohol as they were directed they used a sample so watered that the varnish turned into a magma and they could not get it off without injuring the speculum.

He then asked what you were doing about the Moon. So I told him of her appearance in the new telescope. Pointing to a photograph on the wall, there said he is the best that I could ever do. It is a photograph about 36 inches in diameter, very inferior to the large ones in Mr. Palmer's hall. He then referred to those that you and Rutherfurd[3] had sent him. Of the latter he remarked, he must have had a most superb night. I have made many thousand photographs but never could get one like that and I dont believe if Mr. Rutherfurd makes thousands more that he will ever get such another. So I related to him the anecdote of our being there when Rutherfurd got it. He remarked that the photographic printing of it was very bad and was surprised that Rutherfurd did not have it executed better. Then on a sudden he said, do tell me is he a very rich man. I am told that he has more than £50,000 a year.

He then made a great many inquiries about the construction of the new telescope. I told him that though you had a great many grinding and polishing machines you had come to the conclusion that they all gave inferior results to hand working. That said he is

[3] Lewis Morris Rutherfurd (1816–92) was, like the Drapers, a New Yorker notable for work on telescopes and diffraction gratings.

the conclusion to which I too have come even after having put up a steam engine and all Sorts of devices for the purpose. You cannot prevent machinery giving the mirror *a pattern*. He told me that he had learned from Dr. [Martin ?] who had done all Foucault's[4] work the little secrets and peculiarities of the French method and that they too had come to the Same conclusion. He asked if you were familiar with Foucault's method of examination. So I told him that you had since 1860 been constantly using it and had tested hundreds of mirrors in that way.

But what surprised me exceedingly was to find that with all this experience and advantages he had no conception of the doubling of mirrors or of their oblique action. So I said nothing whatever about it, mentally concluding that if I should find other astronomers here unacquainted with it I would recommend Dan[5] to prepare a memoir on it and send it to the Astronomical Society.

You know that Congress gave $30,000 to a committee for the purpose of making observations on the coming solar eclipse and that [Benjamin] Peirce and several other Americans are now on their way to Southern Europe for that purpose. A Committee of the Royal and Astronomical Societies had applied to the government here for a ship and £1000 to be used for that purpose, this was last Summer. *They were refused.* You can have no idea how great is the mortification here now that the Americans have arrived with their £6000 and no ships. To add to the mortification the American Committee offered to take the English one in charge and convey them to their destination. Thereupon there was such a movement in the newspapers that the English government had to come down with £2000 and a ship. It is amusing to hear the scientific people here explaining the matter away, but it is impossible for them to hide their vexation.

I have seen Trübner and Low and Beli and Dalby.[6] Among them I think I shall be able to have a reprint of the Civil War made. So you see I have not been very idle but on the contrary quite active since coming here. To day I went to the Observatory at Kew and obtained a pretty clear view of every thing. I have written to Dan a long account of the place and its management. The Atheneum Club has sent me an invitation to avail myself of

[4] Jean Bernard Léon Foucault (1819–68), a French experimental physicist, especially noted for work in optics.
[5] Another son, also an astronomer.
[6] English publishers.

their institution. On Friday, day after tomorrow I am to dine there with some notabilities. . . .

(Thursday, Nov. 24. 6 P.M.) Aunty has taken us to day on a visit to Camberwell to see the house in which your grandfather lived at the time we came to America. The place had changed very much, isolated terraces having been built up into continuous streets but still we recognized the house and its surroundings without any difficulty. Though 40 years have elapsed it is still very pretty. Even extravagant New York would call it so. But when I looked on the window at which sweet Mother used to sit with her long black ringlets looking for me coming home it made me very sad indeed. Aunty wanted to go into the house and ask the people to let her look at it, but for me, I had seen enough. Of those who used to be there, your grandmother, mother and Aunt Sarah are dead.

I could not help remarking that I am not quite so strong as I used to be in those old times. I could then *walk* without fatigue, but it tires me to *ride* that distance now. Perhaps it was that mother was there, a bright young girl to welcome me that lightened the toil.

I have not felt very well for the last few days. I sleep very badly being commonly awake from 12 until 5 and then sleeping in a morbidly profound way until the servant wakes me at 8. We all think that the smoky atmosphere of London affects us and that we have been spoilt by being at Hastings . . .

Your Affectionate Father

[no signature]

John William Draper to Henry Draper

Thursday Dec 1st 1870

Dear Henry.

I went last Friday, now nearly a week ago to dinner with Tyndall,[7] and some friends, among them Herbert Spencer,[8] at the Atheneum. It was a very pretty affair and good as the dinner was the conversation was still better. I have also seen Blake and Carpenter, with difficulty getting away from each of them, they had so much to say. I must tell you the details when I get home they would consume too much space now.

Yesterday I went to lunch with Tyndall at the Royal Institution.

[7] John Tyndall (1820–93), Faraday's successor at the Royal Institution. A physicist and a notable popularizer of science.

[8] The English philosopher with whom Draper shared so many views.

He had quite a pretty entertainment. After it we went down into the laboratory, that in which Davy and Faraday made their great discoveries and then he gave me an entertainment far surpassing the lunch. On a very splendid scale he shewed me with one of Duboses electric lamps the recent discoveries respecting the decomposition of Iodide of Allyl nitrite of Amyl etc. by light illustrating *magnificently* the cause of the blue color of the Sky. He shewed me too the effect of the air of respiration and expiration. He had at first invited me for Tuesday and then sent to beg me to postpone it until the next day. I suspected what turned out to be true that he could not get his experiments ready in time and certainly they were very perfect.

He has many great advantages over us. As a proof of this, the Nicol polarizing prism which he used had a diameter of more than 2½ inches and he had two assistants who performed the experiments and they were very skillful men. He simply stood by and directed. I could not help thinking how much more I should have done if I had had such advantages!

In the evening I went by invitation to dine with the Royal Society. The dinner was in the rooms in which Almacks balls, the fashionable balls are given. The dinner was of course very stylish as you would expect, the persons present the first in the land. Among a great many lords and earls we had the Prime Minister Mr. Gladstone and the Lord Chancellor who both made speeches. Previously, I was introduced to them, and received with the utmost kindness. About 250 persons were at the table and it so happened that my lot was cast at table with a very pleasant group, Mr. de La Rue, Mr. Lascalles and Mr. Lockyer.[9] We had indeed a good time. I took quite a fancy to Lockyer. He is such another young fellow as you and has a wife [so Mr. de La Rue told me][10] who takes the same interest in his pursuits that Anne does in yours. He told me that he is coming to America year after next and I am truly glad that you should have the opportunity of seeing him. He is the *rising English Astronomer*. The government has given him $10000 to go out on the eclipse expedition. Mr. Lascalles asked very kindly about our darling mother, she was with us when he gave us that entertainment at Liverpool and he seemed really afflicted to hear that she was gone.

[9] Probably William Lassell (1799–1880), an English astronomer. Joseph Norman Lockyer (1836–1920) had a distinguished career in astronomy.
[10] Draper's insertion.

From the Conversation of these astronomers I found that one of the Australian mirrors was imperfect and will have to be sent back to Grubb, the other was spoilt by varnishing it. The former gave an image of a 1st magnitude Star outside the focus as a line, inside as another line at right angles and at the focus as a X. They spoke not without some reprehension of Grubb, that he has the instincts of a Mechanic, making pretense to all sorts of secrets which he cannot divulge. Lockyer goes next Tuesday to the South of Italy on the eclipse expedition.

This morning Prof Stokes[11] from the University of Cambridge called on me. He gave me some valuable information about the use of quartz and told me that no one has as yet done anything with stellar photography as yet but that Huggins is preparing for it. What he said to me is too long to repeat, so I must reserve it until I get home.

I was at Carpenter's[12] the other morning and had the greatest difficulty in escaping from him.

The pains he took to show me all sorts of new things which he has been dredging from the bottom of the Mediterranean the last Summer almost overwhelmed me. I have declined several private invitations to dinner for this and the coming week for I see it will not do for me, these people would soon make me sick . . .

Your affectionate Father

[no signature]

A. J. Ångström[3] to Henry Draper

Upsala 21 Febr. 1874.

Professor H. Draper

Dear Sir,

Accept my most hearty thanks for the memoir and the Spectrum. It is extraordinarily beautiful and the most perfect I have ever seen. I take the liberty, however, of making a little remark concerning the calculation of the wave-lengths. In my researches on the Solar-Spectrum, I have also made use of coincidences of lines in different Spectra to control the correctness of the measurements, but if the coincidences were observed only on one side of the middle

[11] Sir George Gabriel Stokes (1819–1903), an English physicist and mathematician.

[12] Probably William B. Carpenter (1813–85), the naturalist-oceanographer.

[13] Anders Jonas Ångström (1814–74), a Swedish physicist and astronomer, very active in spectroscopy.

position, I found that errors would easily arise through a little displacement of the different Spectra. You have the kindness to say in your letter that you should like to send me a *glass negative,* if I only could indicate any way in which it could be done. I think that very easy. If you only put it carefully in a small box, I am sure it can be sent by mail and will certainly reach me. The transport will of course be paid at the arrival.

I have the intension to publish a second memoir on the "Spectre solaire" and will therein treat also the Ultra-violet part of Spectrum, and then it would be of very great importance to have resource of your eminent photography. I do not think myself capable of producing a photography as perfect as yours, but the values of the fixed points I shall possibly be in opportunity to determine with sufficient exactness.

I enclose a letter to Mr. Rutherfurd whose direction I do not know, in the hope that you will have the goodness to send it to him.

In the letter I ask for a grating of his making and in case he will grant my request it could possibly be put in the same box as your glass-negative.

Yours truly
And. Jon. Ångström

Henry Draper to Ångström

March 18th 1874

Dear Sir

I am very much obliged to you for the kind expressions you use in relation to my photograph of the diffraction spectrum. They are doubly agreeable because you are the greatest living authority on such matters. I shall examine carefully with a Rutherfurd grating as soon as an opportunity offers the matter of relative displacement of spectra which you mention and I wish to express my thanks for that criticism. For the present however all my time must be given to arrangements for photographing the Transit of Venus. The United States Government Commission has invited me to take control of the photographic organization of 8 parties to be sent to the northern and southern hemispheres and I have of course acceded to so flattering a request.

In regard to sending you a glass negative the difficulty has been that our government does not allow glass to be sent through the mails but I will have a box made and send it by express (transit paid).

I will watch with much pleasure for your second memoir on the "Spectre solaire," the first was a great boon to the scientific world and the same will doubtless be true of the second.

Mr. Rutherfurd is at present in Paris and a letter addressed to *Hottinguer & Co Bankers Paris* will reach him. He will be much pleased to give you a grating for he is a gentleman of leisure of great Scientific attainments and who devotes his time to the pursuit of Science. I have had your letter sent to his agent here.

Yours truly
Henry Draper

Will you be good enough to thank Prof. Thalen[14] for his memoirs which arrived safely.

Ångström to Henry Draper

Upsala the 16 May 1874

Dear Sir,

The negative of the Solar-Spectrum, which you so kindly promised to send me, is now arrived *quite safe*. It has given me the greatest pleasure to receive this valuable gift and I hasten to thank you most sincerely.

The negative is extraordinarily beautiful. I congratulate you heartily upon having so well suceeded to produce on one plate only the whole violet and ultraviolet Spectrum. The mass of various occupations, crowding on me at the end of the University-term make it, however, impossible for me to undertake before June any scientific labours and thus I am obliged to put off until then a more detailed study of Your Spectrum. Accept once more my most cordial thanks and believe me, my dear Sir with the most profound esteem

Yours truly
And. Jon. Ångström

[14] Robert Thalen (1827–1905), a Swedish physicist and collaborator of Ångström.

Thomas Edison[15] to Henry Draper

Menlo Park N.J. Aug. 3 1877

Prof Draper.

When you were here I did not understand about your discovery of oxygen in the sun. I have just received my Sillimans Journal and I now see its great importance, its both a discovery of oxygen in the sun and in Spectrum Analysis. I notice that the iron lines in the air spectrum are slanting while the O lines are perpendicular also that in my photo the two spectrums have not been matched the lower or air spectrum being a shade too far to the right, I also notice two dark lines in the large N line next to Al. There appears to be a white line between the two Al lines which is composed of dots. I suppose these are due to imperfections of the photo. Since you were here I have perfected the Speaking Telegraph so that it is now absolutely perfect and talks as plain as you and I do, the barking of a dog, jingling of keys clapping hands, etc. comes perfect. I have found one more phenomenon with the scintillations if two similar metals are used the scintillations go in all directions by for instance platina and Al are used the Al scintillations shoot out horizontal and below angle of 45° while platina scintillations shoot upwards between 45 and 90. I have already 11 Characteristic Scintillations. When you are down this way again call and see me.

Yours

Thos. A. Edison

Edison to Henry Draper

Menlo Park, N. J. Aug. 8 1877

Friend Draper:

After receiving your letter and studying your drawing a little I noticed that the aluminium flame in your sketch darts out sidewise like iron but not so much, on referring to your photo I notice that the two aluminium lines are also slanting as well as the iron but not nearly so great about in the proportion shewn in your sketch. I have studied your oxygen coincidences and if anything is certain in Spectroscopy you have certainly proved the sun guilty of harboring that Chemical hyena Oxygen. There is one thing about this

[15] Edison was always an avid reader of scientific journals. Some of his work (i.e., the Edison effect) was pure physics. In 1878 Edison accompanied Draper on an expedition (to Wyoming) to observe a total solar eclipse. Edison devised a special instrument for the party.

spectral analysis that I cannot get through my head and that is, why does each Elementary substance give a number of lines of say different vibrating times. The particle which gives one line cannot to my mind be like in every particular with the particle that gives a different line. The only way I can figure it out is that there is but one elementary atom, and the aggregate together to form the so called Elementary metal or substance for instance iron is composed say of 490 atoms which form a whole, sodium 2 atoms etc: and these great number of lines is due to the break up of a molecule: Each taking from the molecule a different vibrating line or a harmonic of the fundamental time of the molecule. When everything that grows or lives can be made of 3 or four substances, as inorganics I dont see the wisdom of having so many elementary substances in the inorganics. I will try and call at your place and see how you peek at the almighty through a Keyhole.

Yours

T. A. Edison

P.S. If the Spectra lines of all metals slant and non metals don't it might be used as a distinction. E

Henry Draper Papers, New York Public Library

Henry Augustus Rowland

In 1871 a young man at Rensselaer Polytechnic Institute at Troy, New York, sent the first of a series of three papers to *Silliman's Journal*. Henry Augustus Rowland (1848–1901) already had a few minor publications to his credit. A graduate of RPI, who had briefly taught in Ohio, he had returned to RPI as a faculty member and his goal was to become a physicist. This, his first serious paper, dealt with magnetism. Rowland intended it and its two successors to be his passports to the world of research.

By the narrowest of margins, Rowland avoided a crushing setback at the outset of his career. His first paper was sent to Wolcott Gibbs at Harvard for appraisal. James Dwight Dana, the editor of the *Journal*, first accepted the paper and then rejected it, an opinion Rowland concurred in. The story of the second paper, "On Magnetic Permeability and the Maximum of Magnetisms of Iron, Steel, and Nickel," is told in the letters that follow.

It was flatly rejected on advice of "two excellent physicists" of New

Haven. Undaunted, Rowland had the courage to submit it to the great English physicist, James Clerk Maxwell (1837–79), *the* authority on electricity and magnetism in that period. Perceiving the merit of the article, Maxwell had it published and encouraged Rowland to continue. After its acceptance by a British journal, Dana offered to publish the third article simultaneously but had to withdraw the offer for lack of space in the proper issue. Assured of outlets for his articles in the leading journals in Britain, Rowland could now afford to view his earlier rejection with philosophic calm.

But Rowland lacked a suitable university post. RPI was not a good base for a man interested in experimental physics. Early in his correspondence with Yale scientists, the latter advised him to come to New Haven to get a Ph.D. (Yale was the first American university—1861—to establish a continuing graduate program leading to the doctorate.) Rowland resisted that lure. When Daniel Coit Gilman (1831–1908) was organizing The Johns Hopkins University in Baltimore, avowedly as a model of what graduate education should be, he heard of the scientist in Troy, asked him to become one of the original faculty and, when Rowland accepted, sent him overseas to study.

Johns Hopkins was an immediate success. Its output of trained men, soon reinforced by products of other graduate schools, was too much for the old, already feeble tradition of amateur and nonspecialized activity. Concurrently, professional societies and professional journals rose to overshadow the institutions of an earlier day. When Rowland's colleague at Hopkins, Ira Remsen (1846–1927), an organic chemist with a doctorate from Germany, found that *Silliman's Journal* would not publish the work of his students, he founded a new periodical, the *American Chemical Journal,* in 1879. Dana was forced to write to Gilman (an old associate from Yale) in 1884 (see p. 274*f.*) asking that no more new journals be founded and assuring him that Rowland's students would get a fair hearing in *Silliman's.*

Rowland had a very distinguished career as an experimental physicist. Soon after coming to Baltimore he turned to optics, the specialty in which he made his greatest contribution, continuing what was almost an American specialty in the latter decades of the past century, the analysis of light by the use of the diffraction grating. In 1868 Rutherfurd in New York had succeeded in ruling two-inch-long gratings on speculum metal with about 20,000 lines. Henry Draper had used a Rutherfurd grating to study the spectra of the elements. The lines in the grating are very fine, closely spaced parallel grooves which spread a beam of light and permit analysis of the wave lengths of the constituents of the light. The resolving power of the grating depends on the number of lines, and it is essential that the grooves be identically shaped, equidistant, and exactly parallel. Rowland devised a more accurate machine for ruling the grooves and was the first to produce diffraction gratings on concave metal surfaces. The resulting gratings had about one hundred thousand lines of six inches in length. The Rowland grating remained the height of precision instrumentation in its field until surpassed by Michelson's echelon grating. Rowland produced much valuable data confirming existing theories or providing leads for new theories. But he was not a framer of theories.

Rowland is important in the history of the physical sciences in America for what he stood for, basic research. He had no hostility to applied research; after all, his training was in engineering. In the latter years of his life Rowland did work in electrical engineering. But he was primarily interested in the advancement of knowledge, not in inventing. He stood for pure research, and he trained many students who were imbued with his viewpoint. Unlike his two great contemporaries, Michelson and Gibbs, he was not retiring nor reluctant to enter the public fray. As early as 1883 he called for more pure science; shortly before he died Rowland repeated this plea in the address that appears at the end of this book.

Dana to Rowland

New Haven June 20 1873

Dear Sir

Our physicists here do not encourage us in publishing your article. We would insert an abstract of your results occupying 3 or 4 pages.

Yours truly

James D. Dana

Dana to Rowland

New Haven July 2 1873

Dear Sir:

Your paper was submitted to two excellent physicists of this city, one of them profound in the Science;[16] and they agreed in their advice as to publication. They thought that you needed more study before you could handle the subject satisfactorily. They also thought that a brief statement of your experimental results was desirable, and hear our proposition. We should have much preferred to have had advice in your favor if such was in accord with their judgment. Thorough articles in physical subjects from American authors are what we especially desire to publish. We hope that you will continue your studies in the department of physics. Should you desire to consult either of the physicists referred to and will come to New

[16] It is rather odd that Dana did not forward the paper to A. M. Mayer of Stevens Institute, the only physicist on the editorial board. In retrospect the only profound physicist in New Haven in 1873 was Gibbs, but there is no certainty that Dana thought so in 1873 or even sent the paper to him at all. No one can say with assurance who the two were. What they and Dana had in mind was for Rowland to enroll in Yale's graduate school.

Haven next September (when they will be at home) I will take pleasure in introducing you. Regretting your disappointment. I am

Yours truly

James D. Dana

P.S. July 26. This letter has been to Newark N.J., and has been returned by Postmaster.

[Undated note]

When this letter was received July 30 I was in possession of Maxwell's letters informing me of the value of my results and the fact that they were to be published in the Phil. Mag.[17]

H.A.R.

James Clerk Maxwell to Rowland

Glenlair Dalbeattie 9 July 1873

Dear Sir

Your letter and MS have been forwarded to me this morning. I have read your paper with great interest and I think the whole of it is of great value and should be published as soon as possible because the subject is one of great importance, and the value of results such as yours is just beginning to be appreciated.

I am sorry that your paper has arrived too late to be communicated to the Royal Society of London, and there will be no more meetings till November otherwise it would have given me great pleasure if you would allow me to communicate your paper to the Society. I hope however that if you have any other papers of the same sort you will give us an opportunity of making them known in this country.

I gather however from your letter that you would consider the Philosophical Magazine a suitable place for your paper to appear in. It is certainly the best medium of publication for any researches in exact science. There are several other scientific periodicals but most of them circulate among a class of readers such that their editors are apt to be suspicious of any article involving exact methods.

Your M.S.S. is so clear both in style and penmanship that I think the correction of the proofs will be a very easy matter. If I sent

[17] Rowland sent the paper to Maxwell upon receipt of the June 20 letter without waiting for a fuller letter of explanation.

it to the Phil. Mag., and if it is likely to appear at once, would you rather have the proofs sent to you or will it be sufficient if I look them over?

I may be mistaken in glancing over the paper but I have not seen clearly whether your magnetization formula rests on a purely empirical foundation, or whether you can give any physical reason for adopting it. You have also an investigation about a cylindric magnet. If as I suppose the work is similar to Green's I should like to see it in full, for if you have satisfactorily got over a piece of Green's[18] work which is very precarious mathematics you deserve great credit.

Yours very truly,
J Clerk Maxwell

Dana to Rowland

New Haven Aug 2 1873

Dear Sir:

I am glad to hear that you have been gratified by a commendation of your paper from a very high source and that it will receive honorable treatment from the Phil. Mag. It was a subject outside of my range; and having asked advice of physicists here I could not do otherwise than as I did. We will publish the abstract you may prepare. But it should be delayed until after the publication of the article abroad. This will not be any objection on our part to inserting the abstract.

If you try us again some time, you may meet with a more satisfactory reply.

Yours truly
James D. Dana

It has for years been my oft-repeated desire that our country should produce some experimental physicists, and I am glad to believe that you are going to add to the very small number now existing among us.

Yours J.D.D.

[18] George Green (1793–1841), an English mathematician.

Rowland to John Trowbridge[19]

Troy Feb. 1 1874

Dear Sir,

I see in the Amer. Jour. of Science for July 1874 p. 21 a short article by David Sears on the Magnetism of soft iron in which he uses a method which I have long ago used in 1870–1 by which I have obtained a very large number of results which I expect soon to publish. Hence I would like to know whether Mr. Sears claims that method: I ask this because he may have obtained it from a paper of mine which I sent to the Amer. Jour. in 1871 containing a description of my experiments and which after laying around for several months and being at Harvard for a long time was at last rejected by the same person [Wolcott Gibbs] who recently rejected my paper on Magnetic Permeability, and so was not published. I do not suppose he did it *knowingly* but only that the idea might have come from there originally and after floating around for sometime without an author been finally seized upon. I have now developed this method into a complete system by which the distribution of magnetism in any case can be determined with great accuracy in absolute measure and this I will soon publish. Hence I would be very much obliged if you would inform me as to this matter.

Yours Truly
Henry A. Rowland

Maxwell to Rowland

Glenlair Dalbeattie 9 July 1874

Dear Sir

It is only within the last few days that I have got an opportunity to go over your paper. I have sent it to the Philosophical Magazine as I think it will be most valuable as a continuation of the former paper. I should mention that Table V is omitted. It should contain an account of magnetic cobalt *cold*. If you can send it to me or better to the Editor of the Phil. Mag. Red Lion Court Fleet Street London E. C. it will make the paper complete.

I hope you will be able soon to prepare a paper on this or some kindred subject for the Royal Society. Of course such a paper should contain enough of explanation to render it intelligible without refer-

[19] Trowbridge (1843–1923), then an assistant professor of physics at Harvard.

ence to former papers though you may quote former papers for facts and results.

Prof. Stoletow[20] was at Cambridge in June. He is to have a laboratory built for him at Moscow and he took great interest in arrangements of the Cavendish Laboratory. He had read your researches on magnetism of course and he seemed to wish to go on with the subject himself. He is quite a young man, he speaks excellent English and is full of energy in a small compass.

We have hardly had time to do any experiments on magnetization at the Cavendish Laboratory but some of the effects of successive magnetizations are very remarkable. They maybe seen either by the method of induction which gives the differences, or by the direct action of the magnetized wire on a needle, which is in equilibrium between the effects of two coils traversed by the same current.

We have the large coils half a metre diameter belonging to the British Association and we have been determining the constants of our galvanometers by putting the galvanometer in the centre of the great coils and finding the current which gives a null effect.

We have also been comparing the induction of one of the great coils on the other (at ¼ metre distance) with that of the inner coil on the outer coil of an induction apparatus.

When the resistances are so arranged that there is no induction "kick", there is still an induction "jerk".

We have first, thus to get rid of the direct "shove" which arises from the direct action of the coils on the galvanometer needle. Then we get rid as far as possible of the "kick" which measures the quantity of the transient current. There remains the jerk which measures the time integral of the transient current

$$\text{Shove} = 2\gamma \qquad \text{Kick} = \int\gamma \qquad \text{dt Jerk} = \int\int\gamma\,\text{dtdt}\,.$$

To eliminate the jerk we make the needle swing through an arc of double the jerk then when the needle is at rest at the end of the swing we make or break contact which jerks the needle into its position of equilibrium and leaves it there at rest if there is no kick left.

Yours truly

J Clerk Maxwell

[20] Probably A. G. Stolotov (1839–96) of the University of Moscow.

Rowland to Maxwell

Graz March 1876

My Dear Professor Maxwell,

Your last letter, which reached me in Dresden, was read with the greatest pleasure, and I could not but admire the beautiful expedient by which you avoided the error due to heating. While in Göttingen, I had the pleasure of seeing the apparatus used by Gauss and Weber and also that more recently used by Kohlrau[s]ch[21] in the determination of the absolute value of Siemen's unit.[22] It seems to me that the form of the earth inductor was such that it would be impossible to find its area with accuracy. So far it seems to me that the *accurate* measurements of resistance either absolutely or relatively is an English science almost unknown in Germany.

The number of new laboratories for Physics, Chemistry, Physiology etc. which there are in Germany and the number to be built is quite astonishing. However, the size of the buildings is quite astonishing. However, the size of the buildings is quite deceptive seeing that the professors and often their assistants live in the building in most cases! Today I have seen the splendid new physical laboratory here and was, on the whole, quite well pleased with it, I would have rather seen more instruments for use and fewer for illustration: I am surprised to find the instruments of research used here often quite poor. The observatories I have visited are very far inferior to many in America, and yet the work done in them is superior in quantity and *perhaps* in quality to most of it in our country. I am doing my best to study the cause of the great amount of work done in Germany, and the low state of the higher education in America. It is only since coming to Europe that I have been able to understand my own countrymen and appreciate their good qualities. In the face of all the evil reports which come from home I believe I can say with pride that there is not a more moral people on earth than our own, and this will account for some of our social habits which I often see criticized. Prof. Sylvester of London is to be our professor of Mathematics in Baltimore, and I am delighted to believe that our University will take a higher stand with respect to education than any other in America, and will encourage *research* in all departments of learning. As there cannot be much done for several years, I hope to have time

[21] F. W. G. Kohlrausch (1840–1910), a German physicist best known for his work in electricity and magnetism.

[22] A unit of electrical resistance.

to prepare myself better. I sail for home on May 6th and shall be in Cambridge sometime in the last of April when I hope very much to see you and Mrs. Maxwell, whose health I hope has improved.

Yours Very Truly
Henry A. Rowland

Rowland to D. C. Gilman

Paris April 20th 1876

My Dear Professor Gilman,

On arriving in Paris after my journey through Germany, I find your two letters of March 30 and 31 awaiting me. You are perfectly right in appointing young men for only limited periods. If my re-appointment is to depend upon the amount and quality of the scientific work which I do within the next five years I am content: but it is to be understood that I do not profess to be able to teach the A. B. C's of physics as well as hundreds of other persons that you can find.

As to the title to be chosen it is hard for me to choose. The subjects which are at present being most investigated, are Electricity and Magnetism, Heat, and Acoustics, in the order I have named. Should I take Elec. Mag. and Heat, it would leave only Optics and Acoustics for another man and it would be hard to find one who is willing to take these.

To the terms Experimental and Mathematical Physics I object since all physics must be mathematical and experimental.

[not signed]

Rowland Papers, Johns Hopkins University

Trowbridge to Gilman

Leamington, England Nov. 12, 1882

My dear Mr. Gilman:

I am very late in acknowledging your kind telegram announcing Rowland's departure from Baltimore. He arrived in Paris about one hour before me. Without concerted action we went to the same hotel. Thursday morning Oct. 26 we went to the American Legation and the American Minister appeared to be much relieved in mind when he saw us; for someone blundered in Paris or in Washington. The Congress met on the 16th of October and was about to adjourn or rather dissolve; but owing to the solicitations of Mr. Morton,[23] they adjourned their last meeting until our arrival. President Grévy's final reception had also been postponed until our coming. We spent Thursday morning in traveling from place to place to find Cochery, the Minister of Postes and Telegraphs to ascertain what the Congress had done. We [did] much sightseeing but accomplished little for Mr. Cochery had just left each place shortly before we arrived at it. Finally we found him and he rapidly read us the minutes of the meeting and informed us that President Grévy would hold a final reception at 3.30 PM and that the Congress would assemble for the last time immediately after the reception.

At the time appointed we drove to the Palace D'Elysées and found the other delegates assembled in one of the State apartments. A chamberlain formed us in line. We stood beside the delegates from England Sir W. Thomson and the American Minister with his interpreter stood beside us also. President Grévy accompanied by an aid de camp in brilliant uniform entered soon and proceeded down the file of delegates, bowing to each of us as we were introduced. The one military man was a queer survival of past splendor. Mr. Grévy looked very stiff and phlegmatic. Having gone the round he delivered a short speech in which he spoke of the magnificent strides of scientists and the pride of France in having gathered such a body of men in her capital. When he had concluded a very frenchified man, possibly speaker of the Corps Legislatif, sprang into the arena from somewhere, no one knew where and read us a long address. At its close we went into an adjoining room, filled with great pictures and glittering with gilt ornaments and the President shook us each by the hand, offering me his left hand which I took by one finger. Then the delegates drove rapidly to the foreign office where the final meeting of the Congress took place. Cochery presided and the results of the meetings were read. The conclusion was that no final action could be taken until further experiments had been made to establish the value of the ohm and the congress adjourned with the hope of meeting again next year.

After the meeting Mascart, Sir W. Thomson, Wiedemann, Kohlrausch and Rossetti went to Rowland's room to see his gratings; they were dumbfounded. Mascart kept repeating with bated breath

[23] The American Minister to Paris. The occasion was an international electrical congress.

"Il faut qu'on commenserat encore" or words to that effect.[24] We dined at the American Ministers Mon. eve. Oct. 30 and left for London Tues. Oct. 31.

In London we have met Abney, Lockyer, Guthrie, Clifton, Preece, Hughes, Cary Foster, Adams (of Leverrier fame)[25] and many others. Rowland read a paper on gratings before the Physical Society and the enthusiasm was tremendous. Johns Hopkins University was lauded to the skies. Rowland has already made himself immortal. I felt very patriotic as I listened to him and saw the astonishment of the Englishmen. Our visit has been very profitable in various ways.

I have seen the laboratories of England and have learned something. There is likely to be a large delegation of Englishmen at the Montreal meeting of the B A association [British Association for the Advancement of Science] in '84. Please remember me to Mrs. G. and Misses G. I have the pleasantest recollections of last summer spent with you all.

Very truly yours,
John Trowbridge

P.S. We dined with Minister [James Russell] Lowell last Friday evening and we are now on our way to Liverpool intending to sail on the Britannic White Star on the 16th.

Trowbridge to Gilman

Cambridge Nov. 30, 1882

My dear Mr. Gilman:

I wrote to you from Leamington, England, giving an account of Rowland's great triumphant march and I wonder at your not receiving the letter. I hope that it is not lost for it contained my acknowledgement of your kindness in telegraphing to me on the eve

[24] E. E. Mascart (1837–1908), a French physicist engaged in the photographic mapping of spectra; Sir W. Thomson, later Lord Kelvin (1824–1907), one of the great physicists of the last century with many contributions in both theoretical and applied work; G. H. Wiedemann (1826–99) worked in electricity and magnetism and edited one of the leading scientific journals in Germany; F. Rossetti (1833–95) was a professor of physics at the University of Padua.

[25] Sir W. deW. Abney (1843–1920) was doing research on the solar spectrum; Frederick Guthrie (1833–86), a chemist; R. B. Clifton (1836–1908) was professor of experimental physics at Oxford; Hughes is probably D. E. Hughes (1831–1900), who had taught in America and was doing both applied and theoretical work in electricity and magnetism; Sir William H. Preece (1834–1913) was an electrical engineer; George Carey Foster (1835–1919) was a physicist at the University of London.

of my departure from Boston, and late enough acknowledgement which our hurried travelling in Europe can only excuse. Rowland invited Mascart, Sir Wm. Thomson, Wiedemann, Rossetti and Kohlrausch to his room at the Hotel Continental in Paris and showed them his photographs and gratings. It is needless to say that they were astonished. Mascart kept muttering "superbe" "magnifique". The Germans spread their palms, looked as if they wished they had ventral fins and tails to express their sentiments. Sir W. Thomson evidently knew very little about this subject and maintained a wholesome reticence, but looked his admiration for he knows a good thing when he sees it and also had the look that he could inform himself upon the whole subject in fifteen minutes, when he got back to Glasgow. We left the French capitol with the feeling that there was little to be learned there in the way of physical science, and having sent for the above scientists as heralds to proclaim the preeminence of American diffraction gratings and in applied electricity (having also dined at the American Ministers in Paris) left for London. In England Rowland's success was better appreciated if possible than in Paris. He read a paper before the very full meeting of the Physical Society. De La Rue, Prof. Dewar of Cambridge,[26] Prof. Clifton of Oxford, Prof. Adams (of Leverrier fame), Prof. Cary Foster, Hilger the optician, Prof. Guthrie, and other noted men being present. I was delighted to see his success. The English men of science were actually dumbfounded. Rowland spoke extremely well, for he was full of his subject and his dry humor was much appreciated by his English audience. When he said that he could do as much in an hour as had hitherto been accomplished in three years, there was a sigh of astonishment and then cries of "Hear! Hear!" Professor Dewar arose and said "We have heard from Professor Rowland that he can do as much in an hour as has been done hitherto in three years. I struggle with a very mixed feeling of elation and depression. Elation for the wonderful gain to science and depression for myself, for I have been at work for three years in mapping the ultra violet." De La Rue asked how many lines to the inch could be ruled by Rowland. The latter replied "I have ruled 43,000 to the inch and I can rule one million to the inch, but what would be the use, no one would ever know that I had really done it." Laughter greeted this sally. This young American was like the Yosemite, Niagara, Pullman parlor car; far ahead of anything in England. Professor Clifton referred in glowing terms

[26] Sir James Dewar (1824–1923), a chemist specializing in low temperature phenomena; Adam Hilger was a scientific instrument maker.

to the wonderful instrument that had been put into the hands of physicists and spoke of the beautiful geometrical demonstrations by Rowland. Professor Dewar said that Johns Hopkins University had done great things for science and that greater achievements would be expected from it. Captain Abney wrote a letter (he having started on his inspection tour to various science schools) which Rowland ought to show you, for after having been read at this meeting, it was given to him. Lockyer before he saw Rowland's results reminded me of this [crowing cock drawn] and afterwards of this [drooping cock drawn]. Lockyer evidently is much disturbed and was not at the meeting of the Physical Society. He does not impress me as a gentleman. Did not call on us in London. I was charmed with Abney. He is evidently of good family and could appreciate Rowland's generosity. He kept saying to me "Rowland is too generous" having been the recipient of one of the best gratings.

Introduced Rowland to a Foxhunting gentleman, an old acquaintance of mine and I imagine Rowland got enough of English Foxhunting. On my return one evening from Birmingham, I found him stretched on the bed, a symphony in red and brown mud, his once glossy hat crushed into nothingness, his topboots once so very new a mass of Warwickshire mud. He dryly remarked that he "guessed there wouldn't be any trouble about getting his hunting suit through the custom house now!" He came very near breaking his neck, having been thrown on his head before he "could calculate his orbit" as he remarked. I couldn't help shuddering from friendship and from love of science. I am sure that he thoroughly appreciates all you have done for him. I want you to use your influence with him to give me one of his best gratings: I am influenced both by selfishness and by a desire to secure the Rumford Medal for him. If the men in this neighborhood could see his results, I am sure that he could get the medal. I say this in confidence to you. The plans of the Physical Laboratory are almost decided upon.

John Trowbridge

Rowland Papers, Miss Harriette Rowland
Transcript furnished by Johns Hopkins University Library

Dana to Daniel Coit Gilman

New Haven June 20, '84

My dear Mr. Gilman

My reply to your allusion to geology and the Journal was perhaps misleading, and I write to add that the predominance of geology over physics is not our wish. It has always been so, because geological workers are many, and physical very very few. I never open a package of MS and find it geology without some feeling of disappointment. Our covering the department of physics is necessary to the welfare of the Journal in several ways. Geological papers began to come in freely about a year since, thro' the activity of the U. S. Geological Survey. But they never interfere with the publication and early publication of any physical paper that may be received. I hope that Prof. Rowland will continue to favor us, as he has done; we make no objection to simultaneous publication abroad. We have physical papers for each of the August and September numbers. I hope therefore that you will not find it necessary to make or issue a Johns Hopkins Journal of Physics. A purely Geol. J. will not pay here.

Sincerely yours,
James D. Dana

July 12. My letter has been to Newport, and is now here again. Learning that you are in Nantucket I send it off once more. I have had my first installment of vacation at Lakeville, Ct.

Dana Papers, Yale University

Newcomb and Michelson

In 1878 Albert A. Michelson (1852–1931) was an ensign serving in Annapolis in the Department of Physics and Chemistry; Simon Newcomb (1835–1909), the Superintendent of the Navy's Nautical Almanac Office, had an exalted position which would have delighted Gilbert and Sullivan—professor of mathematics in the Navy's Corps of Professors of Mathematics. They had many things in common: the humble origins required by American folklore of its heroes; common membership in a languishing naval scientific tradition derived from the older geographical science; and an interest in the velocity of light. More important were the significant differences, subtle and gross, dividing these two exemplars of different scientific approaches.

Newcomb was a poor boy from Nova Scotia who escaped an unhappy apprenticeship by running away to the United States. With the encouragement of Joseph Henry and others who recognized his talent, he was given a position at the Nautical Almanac Office, then at Cambridge, which enabled him to attend the Lawrence Scientific School at Harvard. Newcomb was an outstanding mathematical astronomer in the classical tradition. As a mathematician, he certainly far outstripped Michelson. Newcomb also had ambitions to be a "practical" astronomer, to make observations of stars and to do astrophysical work. Here he was not so successful.

In at least one other crucial respect, Newcomb, the classical mathematical astronomer, differed from Michelson, the specialist in experimental optics. Newcomb was a public figure, writing and lecturing widely on topics far removed from astronomy. He was a renowned economist in his day. Few twentieth-century scientists would resemble Newcomb in his uninhibited ranging over the intellectual landscape. They would be more like Michelson in their absorption in their specialties.

Michelson was born in German Poland. At age two his family came to America, eventually settling in Nevada. When the Congressman from Nevada held an open competition for an appointment to Annapolis, Michelson tied for first place with two others. Passed over for the son of a poor, crippled Civil War veteran, Michelson went East with his Congressman's blessing to get an appointment from President Grant. By some still obscure stretching of the regulations (the Presidential appointments were all filled), Michelson was taken into Annapolis. The traditions of the age of geographical science were still evident in the service academies, which provided a scientific education well above the American average.

Newcomb was particularly interested in determining the solar parallax, the same object that had sent Gilliss on his expedition to Chile. His most strenuous attempts at practical astronomy occurred during the period of his association with Michelson, 1878–85, involving the observations of the transit of Venus of 1882 and the determination of the velocity of light. Both efforts were aimed at determining the solar parallax. While Michelson was not involved in the former, it may have influenced his relations with Newcomb. For some unexplained reasons Newcomb never published his transit of Venus results; perhaps unfairly, his velocity of light determinations are overshadowed by Michelson's. On his return from the transit of Venus expedition there is evidence that Newcomb was depressed, if not suffering from a nervous breakdown. Speculation on these matters decades after the facts is dangerous, but there is the distinct possibility that this distress at least partly stemmed from a recognition of his limitations as an experimentalist which were sharply underlined by the comparison with Michelson.

As early as 1867 Newcomb had suggested the desirability of accurately determining the velocity of light as a means of obtaining a reliable value for the radius of the earth's orbit. At Newcomb's instigation the National Academy of Sciences in 1878 considered his proposed velocity determination. Unknown to him Michelson had perceived a modification of Foucault's revolving mirror method in November of the previous year. By March of 1878 Newcomb learned of Michelson's work. On April 26 Michelson opened a correspondence with Newcomb, some of which is printed below. Michelson's achievement was certainly impressive.

When Michelson published the results of his early work, he mentioned that the revolving mirror cost $10. Some later accounts have picked up this figure and presented us with the romantic image of the young genius obtaining great results on a pittance while Newcomb was bumbling away lavish Congressional appropriations. Actually, the kind of experiments Newcomb and Michelson had in mind were not inexpensive, and in the spring of 1878 both faced the same problem—how to get funds. Michelson solved the problem first by getting $2,000 from his father-in-law. By the end of January of 1879 Michelson was obtaining results. Newcomb's proposed determination was placed in jeopardy because of Michelson's earlier findings. But with the backing of the National Academy and of the Secretary of the Navy, he received a $5,000 appropriation on March 3, 1879. In the fall Michelson was detailed to the Nautical Almanac to assist Newcomb in the experiments. A year later Michelson left for two years of study in Europe. While overseas he resigned from the Navy.

In Europe Michelson started work on what was to be the high point of his career, the detection of the relative motion of the earth and the ether, a fluid then presumed to permeate all space. Michelson was unsuccessful in obtaining funds for his 1881 work from Newcomb, who, as appears from the letters below, did not quite grasp the potentiality of the interferometer invented for the experiment. Funds were provided by Alexander Graham Bell. The negative results of the determinations in Potsdam in 1881 and in the more precise determination by Michelson and Morley in Cleveland in 1887 cast doubt upon some basic assumptions of classical physics.

While Michelson was engaged in this line of research in Europe, Newcomb was reviewing the results of the velocity of light determination. Seeking an exact value for the computation of the dimensions of the universe, Newcomb was embarrassed that his results and Michelson's of 1879 were apart by at least 200 km/sec. Newcomb subjected Michelson's data and his equipment (with the notable exception of the temporarily missing $10 revolving mirror) to a keen critical scrutiny, discovering minor errors which did not resolve the discrepancy between the two results. He finally obtained money from the Bache Fund of the National Academy for a redetermination of the velocity of light at Case Institute where Michelson went on his return from Europe. Eventually Newcomb and Michelson came out with results suspiciously close to one another. With the Bache moneys, Michelson continued his research on the properties of light.

Michelson is often pictured as arising semi-miraculously without any formal training. A more accurate picture emerges if we regard the relationship between Michelson and Newcomb as that of a graduate student to his faculty advisor. This involved initiation in the standards of the discipline; guidance of reading (in 1878 Newcomb sent Michelson a copy of the *Works* of Foucault, his scientific predecessor); and aid and advice in job hunting and fund raising. Like all good teachers Newcomb's influence is apparent even in Michelson's last work on the velocity of light in 1927, where two 1885 suggestions of Newcomb's were utilized. Newcomb also experienced what is both the glory and the vexation of teaching —being surpassed by a pupil.

A. A. Michelson to Simon Newcomb

U.S. Naval Academy Annapolis Md.
April 26th 1878

Professor Newcomb;
Dear Sir,

Having read in the "Tribune" an extract of your paper on a method for finding the velocity of light, and hearing through Capt. Sampson and Capt. Howell [27] that you were interested in my own experiments, I trust I am not taking too great a liberty in laying before you a brief account of what I have done.

The principle of the method is as follows;

S, is a straight-edge from which the light proceeds; *m,* the revolving mirror; *l,* an achromatic lense of long focus, and *M,* a plane mirror. The edge *S* is so placed that its image, formed by the lense, coincides with the plane mirror *M.* With this arrangement, the light retraces its path so as to form an image which coincides with *S,* when the mirror is at rest, or when it revolves slowly. When the motion is rapid the image is permanently deflected.

In the last experiment I made the distance mS, was 15 feet; mM, 500 ft. The deflection was about 0.3 inch.

The distance mS might be considerably increased and the deflection proportionately.

Unfortunately, as I was about to make an accurate observation the mirror flew out of its bearings and broke.

It would give me great pleasure, dear sir, if you could honor me

[27] Capt., later Rear Admiral, William Thomas Sampson (1840–1902) was head of the Department of Physics at Annapolis; Capt., later Rear Admiral, John Adams Howell (1840–1918) was head of the Department of Astronomy and Navigation. Sampson was especially esteemed as a scientist; Howell was an inventor of ordnance devices.

with an interview, in which you could advise me how to arrange some of the details so as to insure good results.

Believe me sir, your obedient Servant
Albert Michelson
Ensign U.S.N.

Newcomb Papers, Library of Congress

Newcomb to Michelson

April 30 [187]8

Dear Sir.

I am very much interested in your experiments on the velocity of light, and greatly obliged to you for communicating the account of them that you did. To have obtained so large a deviation from apparatus so extremely simple, seems to me a triumph, upon which you ought to be most heartily congratulated. So far as I know, it is the first actual experiment of this kind ever made on this side of the Atlantic. I suppose you know that the arrangement of your apparatus and the principle of your method is substantially the same with that adopted by Foucault. The only difference is that your distant mirror should not be flat, but slightly concave, so as to have its center of gravity near the lens.

My own plan was to put the reflecting mirror at a great distance, somewhere from one to three miles, and thus avoid the necessity of an extremely rapid rotation of the revolving mirror. This I think, necessitates receiving the return ray on a different part of the mirror from that which reflects the outgoing ray. I also design placing the revolving mirror in a vacuum. In my own experiments this might not be absolutely necessary; but I regard it as absolutely necessary if any plan is adopted requiring any very rapid motion of the mirror. It is not merely the difficulty of getting a sufficiently rapid motion, but the danger that the image might be disturbed by the vortex produced by the rotation. If a vacuum is used, it seems to me it will be almost absolutely necessary to receive the return ray through a different telescope on a different part of the mirror for reasons which I cannot now stop to explain.

Still, I am not at all sure but that your plan is better than mine. Certainly it is simpler and cheaper; and, therefore, may be worthy of the first trial. Some preliminary experiments may be necessary to decide between them. I cannot, however, go intelligently into any comparison of the relative advantages until I have more exact knowledge of the actual working of your apparatus. I shall therefore content myself with mentioning my general plan for putting the experiment through.

If any one else will make the experiment with all necessary certainty and accuracy, I have no desire to take more than a subordinate part in it. All I want is a result. Professor Barker[28] of Philadelphia advised me that a gentleman of that City contemplates advancing the funds necessary to enable the experiments to be put through. If this is done I shall take no further part in them, than to give such advice and assistance as may seem necessary and to satisfy myself that the result is beyond doubt. If this is not done, I shall make an effort to get an appropriation next winter, sufficient at least, to commence the experiments. If this effort is successful, it would be a question where the experiments should be made. Washington, Annapolis and Philadelphia will all have their advantages, should Annapolis prove to be the most advantageous place, it would be very agreeable to me to have them execute it there. We must not, however, count our chickens so very far in advance of the eggs.

I hope in conclusion that you will get your mirror going once more, and let me know when it is in successful operation, that I may go down and see it with my own eyes.

Yours very truly,
Simon Newcomb

Michelson to Newcomb

Annapolis, Md. Dec. 18th 1878

Dear Sir,

I received a call a few days ago from Prof. Johnson of Annapolis, and was informed that a friend of his, a lawyer, Alexander Evans, has become very much interested in my experiments. He is personally acquainted with the Hon. Alexander H. Stevens,[29] and suggests that that gentleman bring the matter before Congress.

I stated the case to him through Mr. Johnson. The latter has just called and asked me to write to you to see if you would call on Mr. Stevens and inform him about the particulars.

I have been trying to get my apparatus together to start the experiments I intend to make with the $2000 which I have. I have

[28] George F. Barker (1835–1910) of the University of Pennsylvania.

[29] Alexander H. Stephens (1812–83), the former Vice-President of the Confederacy, then serving in the House of Representatives. William Woolsey Johnson (1841–1927), then professor of mathematics at St. John's College. Alexander Evans (1818–88) was a former Whig Congressman from Maryland.

built a small house about 45 ft long and raised about 7 ft from the ground, and erected a brick pier for the distant mirror the whole distance being about 2000 feet. Have ordered from Alvan Clarke[30] a lens 7 inches in diam 160 ft. focal length. I have made a calculation which seems to show that with such a lens I will obtain about thirty times as much light as before (i.e. with the arrangement you saw) and still get a displacement four times as great.

Alvan Clark said there was no necessity for making the lens achromatic, and I believe he is right.

I have been waiting for about three weeks for a Root Blower, with which I expect to get 200 or 250 turns per sec. of the mirror, but it has not yet made its appearance. This is the more aggravating as the dimensions of other apparatus (notably the tuning forks) depends on the speed I am able to obtain.

I have tried forming the image of the slit at the distance of 2000 ft. and it was more distinct than it ever appeared in the Laboratory for I made a number of measurements all of which fell within 0.01 millimeters!

I think the reason of the great unsteadiness on former occasions was the difference in temperature inside and outside the building. I count on obtaining a deflection of 100 or 120 mm hence the displacement can be measured to 1/100 of 1 percent. The distance 2000 ft. could not be measured with much greater accuracy, neither could the speed of the mirror nor the "radius of measurement." I consider therefore that the distance 2000 ft was well chosen. Moreover it is along a level stretch and needs no triangulation.

Another great advantage lies in the fact that about an hour and a half before sunset [the distant mirror] lies almost entirely in shadow.

I intend next Monday to begin experiments using the old lens and bellows. I expect to get results which will not differ more than 1/10 of 1 percent.

Hoping to hear from you soon, I remain

Sincerely Yours
Albert Michelson.

Newcomb to Michelson

December 20th [187]8

Dear Sir,

I have just received yours of the 18th.

[30] Alvan G. Clarke (1832–97), an American instrument maker whose labors greatly facilitated the expansion of astronomy in America.

Last week I wrote the Secretary of the Navy an elaborate letter recommending an appropriation for measuring the velocity of light, which he took to the Senate Committee on appropriations. I fear the Committee did not look upon the subject with favor for they have not inserted it in the Naval Appropriation Bill. Therefore I am a little at a loss what to do next. The only remaining hope is to get it inserted in Sundry Civil Bill either by the House or Senate; and here the influence of Mr. Stevens might be valuable. Can you not get Mr. Evans to write Mr. Stevens on the subject so that if I call on him he will have the matter already in mind. Possibly I may make the next move through the National Academy of Sciences.

Meanwhile I am very glad to hear that you are getting along so well in your own way. When you speak of forming the image of a slit at the distance of 2000 feet, I suppose you mean that you place the distant mirror at that distance. I am also quite sure that Clark is right in there being no necessity for having the lens achromatic. Indeed from what you say it would almost seem that you will attain complete success without the necessity of Congressional appropriation. The weakest point as it now strikes me is turning your mirror by a blower. It is absolutely necessary that your results shall be free from any suspicion of constant error. You must therefore be certain that the velocity of your mirror is absolutely the same during the minute interval between the reflections as during the rest of the revolution. The action on the mirror is intermittent there will always be room to fear that this result is not attained unless you can so arrange it that the action can be applied successively at all points of the circuit of the mirror. In the arrangement which I saw this would be impracticable. Again you must have some way of ascertaining the speed of your tuning fork. If you can rely upon it to a small fraction of one per cent I will be agreeably disappointed. The chronograph wheel seems to me vastly superior for the reason that it leaves no possible doubt.

I shall look with great interest for the results of your experiments with the old apparatus next Monday, and when you have them in good working shape would like to pay you a visit to see them.

Yours Very Truly,
Simon Newcomb

Michelson to Newcomb

Annapolis Md. Dec. 25th 1878

Dear Sir,

Yours duly received. Mr. Evans has already written to Mr. Stevens, explaining matters, and he seems to be favorably impressed.

If by means of a regulator I can obtain steady average pressure with the blower, the fact of its being intermittent would have no other effect than to produce an indistinctness of the image, unless the puffs of air always acted when the revolving mirror was in a certain position; and this to run for half an hour and the seconds mark read to 1/20 of an inch.

However I am quite open to conviction should you still hold to the latter method.

As soon as the weather is warmer I shall measure the *distance* and then shall be ready to make observations.

In a rough trial yesterday, I obtained the following readings

Mirror stationary	Mirror revolving
31.01 millimeters	15.86 millimeters
31.02 "	15.88 "
31.05 "	15.98 "
31.05 "	15.98 "
31.01 "	15.92 "
	15.91 "

Very sincerely yours
A. A. Michelson

Naval Historical Foundation Collection, Library of Congress

Michelson to Newcomb

Annapolis, Md. Dec. 29th [18]78

Dear Sir,

Your last duly received. Many thanks for your kindness; I think I can restrain my impatience till the volumes are bound; and would be much obliged if you could send also the volume of Leverrier's Annals, containing Cornu's[31] paper.

As to the advantage of revolving the mirror in opposite directions, I am not quite clear.

[31] Mari Alfred Cornu (1841–1902), a French physicist best known for his velocity of light determination.

If the fiducial mark is to be measured both times, it seems to me that it is the same as making two separate observations. If the angle between the two deflections is to be measured without measuring the position at rest, it is necessary that the speed of the mirror, for both deflections, be exactly the same, a condition difficult to obtain.

Moreover the apparatus becomes more complicated, provision made for preventing the oil for the bearings from leaving the oils cups etc.

Besides, I can get a displacement of five inches, which I can measure with greater accuracy than the other quantities.

What troubles me most at present is to drive the Blower (which has arrived and has proved satisfactory) with a constant average speed.

I see no simpler way than to use a small steam engine.

Yours very Sincerely
A. A. Michelson

Michelson to Newcomb

Annapolis Md. Jan'y 16th [18]79

Dear Sir,

Yours of the 14th instant duly received. The books arrived in good order. Many thanks for your trouble.

I agree that the account of the experiments is very meager. I was astonished to find that Foucault proposed to extend his distance by means of lenses thus;

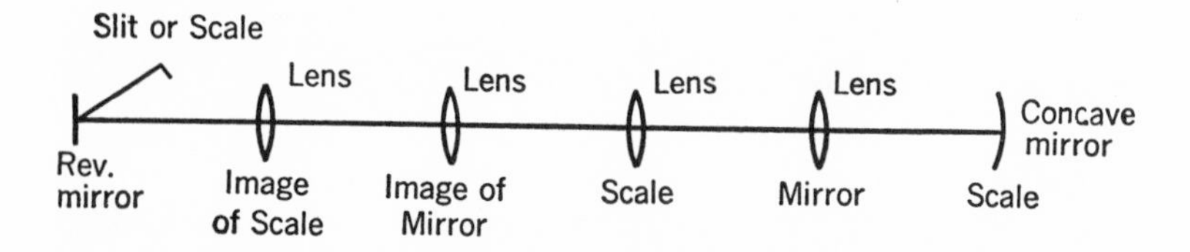

Why he did not carry out the plan instead of using concave mirrors will probably remain a mystery. Had he done so he would in all probability have obtained as great a distance as that which I use.

Clark has written to say that my lens will be ready next week, so that all that remains is to get a fly wheel for my blower, and a bellows for further regulating the blast. The latter is now making, but it seems as though the fates were against my getting the fly wheel, as I have been trying for three weeks to get someone to make it.

I have also ordered a micrometer which dispenses with the plane glass mirror, thus;

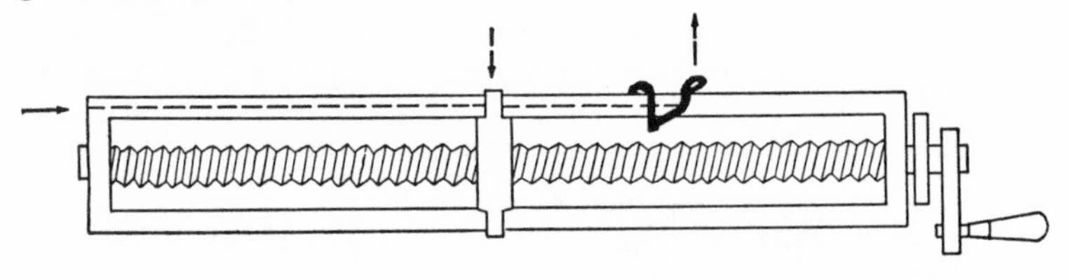

Yours very Sincerely
A A Michelson

Newcomb Papers, Library of Congress

Michelson to Newcomb

Cambridgeport Mass. March 8th 1880

Dear Sir,

The experiments[32] have been delayed on account of the weather. The following contains all the essential points.

First. The calcium light was used at night. The mirror which is the same one which you had made, was sufficiently near to being plane, and was placed at a distance of 180 feet from the revolving mirror. The apparatus was centred and focused and a bright image of the slit obtained. The revolving mirror was then rotated by a cord passed over the axle. The image was bright enough to bisect with ease. It could possibly have been bisected if it were half or third as bright. The slit was narrowed to about 0.005 inch, without apparent diminution in brightness. Apertures of ¾ in. and ½ in. were made and placed in front of the plane mirror. If the former were placed centrally, the illumination should theoretically be almost the same, since this is one half the aperture of the lens.

It was moved about in different positions, and occasionally the light was bright enough to bisect. With the ½ in. aperture it was invisible.

Second. The same mirror was used with sunlight, at a distance of 1400 feet. The error due to the mirror being concave would be small as the mirror had, I believe, a radius of a mile and a half, and can readily be corrected. The "sway" was not very good, so that the adjustment for the return light may not have been perfect.

[32] Michelson is now assisting Newcomb in the latter's determination of the velocity of light.

Owing to the mirror being plane, the image instead of being a slit, was a "star" passing across the field, the path being limited by the size of the slit. In fact it was an image of but a small portion of the slit; thus, (considerably magnified.)

This was bright enough to measure, when the field was darkened by surrounding the apparatus with screws.

The width of the slit was 0.01 inch. This is about the limit of distance with a plane mirror. Possibly if the field had been quite dark this distance could be increased to 2000 feet.

I believe that, as far as is possible, your instructions have been carried out, and in obedience to your orders I shall return to New York. My address will be as before 328 Fifth Ave.

Very respectfully, your obedient Servant
Albert A. Michelson
Master US Navy

Nautical Almanac Records, National Archives

Michelson to A. M. Mayer

Nautical Almanac Office Washington, D. C.
June 26th 1880

My dear Prof. Mayer[33]

The Professorship of Math in the Navy is still vacant and there seems to be no immediate prospect of its being filled. Besides this there is now a law requiring an examination, probably in astronomy and Mathematics and as I make no pretense of being either astronomer or mathematician, I doubt if I could satisfactorily pass such an examination, even if I got the appointment.

In short I have abandoned the said professorship and it therefore behoves me to look about for something else and that p.d.q!

As Newcomb's experiments are underway, and progressing satisfactorily, it may be that in a few months they will be completed, which, unless I was enabled to tender my resignation, would necessitate my departure for Foreign Climes.

To the point, then, if you can do anything possible to assist me in obtaining a fair position in some institution where there is a

[33] Alfred Marshall Mayer (1836–97) of Stevens Institute, a physicist highly regarded in his day.

respectable physical laboratory, and if possible also a respectable salary, you would confer a lasting honor.

I am naturally somewhat anxious about the matter, so, trusting it may receive your earnest consideration, I am

very sincerely yours
Albert A. Michelson

Hyatt-Mayer Collection, Princeton University

Michelson to Newcomb

54 Königgratrer Strasse
Care Frau Dr. Landmann
Berlin Oct 19th/80

Dear Sir,

I do not understand how there could possibly have been so great a change of the zero point as two divisions.

In one of the sets, I remember that I accidentally skipped one number, or else made two observations on the *same* setting, but the number was, I believe, correctly reported. The micrometer did not move, for I guarded against it. In all of my observations except perhaps the first days work, the single wire bisected the space between the other two. In every case the setting was made with the double wires and not with the single ones.

Please send mail etc. to the address given above.

I have not sent my copies of the paper on Velocity of Light to any important persons, so that you need not fear duplication. Am sorry to say that I did not meet any of the gentlemen to whom you were kind enough to give letters of introduction, in either London or Paris, as they were all out of town.

Trusting that with the information in this letter, you may find the observations reliable, I remain

Very truly Yours
Albert A. Michelson

Michelson to Newcomb

Berlin, Nov. 22nd. 1880

Dear Sir,

Your very welcome letter has just been received. It will give me much pleasure to let you know how I am progressing.

At present the work in the laboratory is very elementary, and I am trying to get over that part somewhat hurriedly.

Besides this work I attend the lectures on Theoretical Physics by Dr. Helmholtz, and am studying mathematics and mechanics at home.

I had quite a long conversation with Dr. Helmholtz[34] concerning my proposed method for finding the motion of the earth relative to the ether, and he said he could see no objection to it, except the difficulty of keeping a constant temperature. He said, however, that I had better wait till my return to the U. S. before attempting them, as he doubted if they had the facilities for carrying out such experiments, on account of the necessity of keeping a room at a constant temperature.

With all due respect, however, I think differently, for if the apparatus is surrounded with melting ice, the temperature will be as nearly constant as possible.

There is another and unexpected difficulty, which I fear will necessitate the postponement of the experiments indefinitely, namely, that the necessary funds do not seem to be forthcoming.

Dr. Helmholtz was however quite willing to have me make experiments upon light passing through a narrow aperture, but did not give much encouragement. In his opinion the polarisation arises purely from reflection from the sides of the slit.

The change in colour, he ignores entirely.

With regard to the change in the zero point of the phototacometer, I can only suggest that the position of the observing telescope in its bearings may have changed, as it was not clamped.

With many thanks for your kind interest in my affairs, I remain

Very truly Yours
Albert A. Michelson

Nautical Almanac Records, National Archives

Michelson to Alexander Graham Bell

Heidelberg, Baden, Germany April 17th 1881.

My dear Mr. Bell,

The experiments concerning the relative motion of the earth with respect to the ether have just been brought to a successful termination. The result was however *negative*.

[34] Herman Ludwig Ferdinand Helmholtz (1821–94), one of the great physicists of the nineteenth century.

The apparatus was constructed on the plan which I described in my last letter,[35] and was duly installed at Helmholtz's Physical Laboratory. It was soon found however that the instrument was so extremely sensitive to vibrations that even after midnight, it was not only impossible to make measurements, but the interference fringes were actually invisible.

If this is the case with the instrument constructed expressly to avoid sensitiveness what may we not expect from one made as sensitive as possible! It seems to me that such a one may possibly excel the microphone.

It was thus found impossible to conduct the investigation in Berlin, and accordingly, the apparatus was removed to the "Astrophysicaliches Observatorium" in Potsdam the facilities of which were kindly placed at my disposal by the director, Prof. Vogel.

At this season of the year the supposed motion of the solar system coincides approximately with the motion of the earth around the sun, so that the effect to be o[b]serve[d] was at its maximum, and accordingly if the ether were at rest, the motion of the earth through it should produce a displacement of the interference fringes, of *at least* one tenth the distance between the fringes; a quantity easily measurable. The actual displacement was about one one hundredth, and this, assignable to the errors of experiment.

Thus the question is solved in the negative, showing that the ether in the vicinity of the earth is moving with the earth; a result in direct variance with the generally received theory of aberration.

Of the £100 which you kindly placed at my disposal there are £60 remaining,[36] and as the experiments are now completed this sum awaits your disposition. I have just finished the winter semester under Helmholtz, and on account of the health of Mrs. Michelson, and the children (of which there are now three, the latest arrival, a daughter) I have concluded to pass the summer semester here and will attend the lectures of Quincke and Bunsen.[37]

I presume you may have heard that I have been appointed to the

[35] Not found.

[36] $200 for a great experiment is most reasonable. Times change; today a Michelson would have to negotiate an impressive sum from a public or private bureaucracy.

[37] G. H. Quincke (1834–1924) was a physicist noted, among other things, for research on capillarity. R. W. E. von Bunsen (1811–99) was one of the great names in nineteenth-century chemistry and is best known for work on spectrum analysis.

chair of Physics in the "Case School of Applied Science" in Cleveland. The appointment however is to date from Sept. 1st 1882, and the intervening time I shall probably spend abroad.

Mrs. Michelson joins in sending best regards to yourself and family.

Very sincerely yours
Albert A. Michelson.

N.B. Thanks for your pamphlet on the photophone

Bell Papers, National Geographic Society

Newcomb to Michelson

May 2nd [188]1

My Dear Sir:

I am much pleased to receive your letter of April 17th.[38] I heard a few weeks ago that you had been selected as Instructor of Physics at the Case School of Applied Sciences but had heard no particulars as to time. The Johns Hopkins University also has had it in view to offer you such position as they might have at their disposal but what that position would be I did not inquire as I had some doubt whether you would entertain any offer of the kind.

Your arrangement for getting interference through long distances seems to me very beautiful but I do not see that it can settle the question of the motion of the ether. This motion makes itself felt by a difference in the wave lengths and velocities of propagation in different directions. But when a ray returns on its own path the retardation in one direction is compensated by the acceleration in the other. So that the result is the same as if the ether were at rest, just as in the ordinary measures of the velocity. I have found that, for the same reason, a wave length measured with an instrument in which the ray returns nearly on its own path will give only a mean result. But if the wave length is measured with but a small deflection at the point of displacement as with a transparent ruled plate the effect ought to show itself. This is therefore the crucial experiment I would like to see tried.

I have been much embarrassed at the result I am going to get for the velocity of light. Reduced to a vacuum it will be but little over 299700 [km/sec] and pretty certainly less than 299750 [km/sec].

[38] Not found, presumably like the letter to Bell of that date.

The only point in which there appears to be room for doubt is the distance between the mirrors which was determined by the Coast Survey by a careful triangulation. But on inquiry they assure me that there is no possibility of any error more than the accidental one of one or two decimeters, since the base was measured twice and the triangulations all checked. So it seems that my result will be at least 200 Kilometres less than yours and perhaps more. This leads me to inquire whether you have preserved your micrometer and tape line with which your measures were made. If so I would be greatly pleased if you would loan them to me to investigate their relation independently of any hypothesis respecting the scales of measure. I mention this because here is the only place in which I see any possibility of a systemic error in your results.

Yours very truly,
S. Newcomb

Nautical Almanac Records, National Archives

Bell to Michelson

May 14th 1881

My dear Sir:

I must thank you very much for the letters you have sent me from time to time and if I have not answered them before I can assure you it is from no lack of interest in your work. I think the results you have obtained will prove to be of much importance.

With reference to the balance of the Volta money[39] in your hands I wish you to retain it in furtherance of any experiments you may have in prospect.

I forward copies of my recent paper on the "Production of Sound by Radiant Energy."

Please present one to Helmholtz for me.

Accept my congratulations on the happy addition to your family. With kind regards to Mrs. Michelson and yourself I remain,

Yours truly,
[no signature]

Bell Papers, National Geographic Society

[39] Bell received 50,000 francs from France (the Volta prize) for the invention of the telephone. He used the money to create a laboratory in which work was done on "talking machines." Some of this money was apparently given to Michelson. In 1887 Bell, using the proceeds from his patents on talking machines, founded the Volta Bureau in Washington, D.C., for the increase and diffusion of knowledge on the deaf.

Newcomb to Michelson

June 2nd [188]1

Dear Sir:

When I wrote that the wave lengths in opposite directions would compensate each other I meant considering only quantities of the first order. This agrees with your formula where the difference is of the order v^2/V^2 amounting to 1/50,000,000 I suppose this difference would not be made appreciable by any experimental process, but it appears by your device to be made so, theoretically at least. Still I cannot feel sure but what some little action may come in to nullify the effect of so minute a cause and it seems to me we ought to be able to devise some way of getting a result which would depend only on quantities of the first order. The only way I know of to avoid this is to measure wave lengths by an ordinary ruled transparent plate and I am surprised that no one has undertaken and published a decisive experiment of this kind.

I have written to President Morton[40] to borrow your micrometer.

Yours very truly
Simon Newcomb

Newcomb to Michelson

June 23rd [188]1

Dear Sir:

I have made such examination of your micrometer and tape line as to show that the discrepancy between our results cannot arise from any error in your standards of measurement the micrometer and the tape line. My results for your micrometer are in substantial agreement with your own. The tape line in a rough comparison with the standard meter scale gives it 1/10000 too short instead of too long as you had it. This however may arise from some personal equation in setting upon the divisions of the tape line.

I have not yet investigated the stretch of the tape line as I did not suppose that it could amount to enough to make up the difference.

If I remember rightly the table which supported the micrometer stood on the floor and not upon a separate pier. In this case I should think there might be a small error arising from the spring of the floor but I do not know how it could systematically affect your re-

[40] Henry Morton (1836–1902), President of Stevens Institute in Hoboken.

sults. But there is one point which you do not mention in your paper. In none of your measures could you personally have examined the two ends of the tape line or the two points of measurement at the same time. You must I suppose have had an assistant who brought one end of the tape in juxtaposition with the face of the mirror while you yourself examined the measurement at the slit, or vice versa. Here the question would come in what part of the tape line did your assistant have against the face of the mirror and how did he manage to make an accurate setting.

Again as to the distances, you say the measurements were made from the forward side of one block to the forward side of the second. From the results you give it would seem that some point at or near the zero point of the tape line was used to measure from while you noted how much larger the distance given at the other or 100 foot end.

Now there is no straight zero division on your tape line but simply a little handle. Now what part of this handle did your assistant have in coincidence at one end while you were observing the deficiency at the other end? It seems to me the more natural method would have been to measure from not the zero point but the first division on the line itself.

Do you know if your pier which supported your revolving mirror is still standing so that I can repeat the measure with a metre bar if it appears necessary.

I am about commencing a repetition of my determination between Fort Whipple and the Washington Monument. I can make a rough triangulation for this distance myself and then check the Coast Survey result.

If it turns out absolutely that there is no mistake in the Coast Survey distance the difference of results will be simply unaccountable. The next step in logical order will be to have the determination repeated with your own apparatus and if possible by yourself at least so far as to make sure that your apparatus will give substantially the same result as before.

The eye piece and slit of your micrometer have not yet turned up but I expect to receive tomorrow an eye piece which may possibly belong to it.

Hoping that you are progressing favorably, I remain

Very truly yours

Simon Newcomb

Since writing the above I have received from Prof. Mayer the eye piece of your micrometer. The moving part of the slit cannot yet be found.

Michelson to Newcomb

Heidelberg July 2nd 1881

Dear Sir,

Yours of the 14 ultimate just received. I was under the impression that the micrometer had been left with all the parts complete and was much surprized to hear that this was not the case.

I believe that I removed the eye piece and the slit, substituting therefore a microscope, and one of Rogers scales, but am almost certain that I did not take the parts from the Institute.

I would suggest a similar arrangement for your measurements; namely, a small microscope in place of the eye piece. In the experiments proper, the cross hair consi[s]ted simply of a fibre of silk stretched across the end of the eye piece; thus;

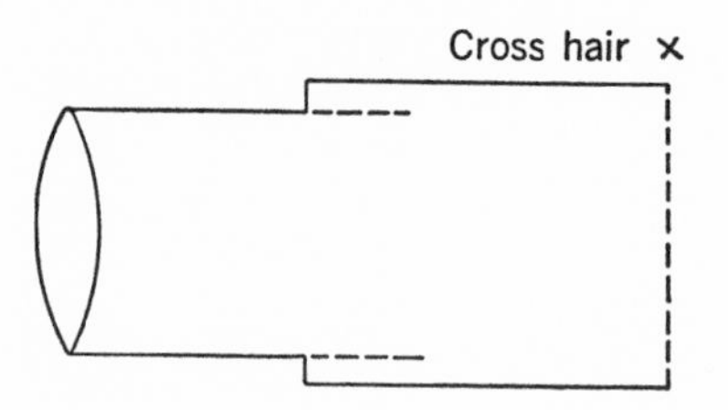

or viewed from the right side, thus;—

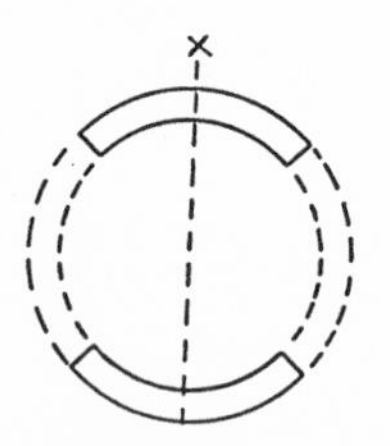

This arrangement was employed in order to have the cross hair as nearly as possible in the same plane with the slit.

I have been thinking over the plan which you suggested for obtaining a derivation of the first order in measuring the relative motion of earth through ether, and the only objection I can offer is that so large a deviation should have been observed before, even without any special arrangement.

For instance, if the angle made by an image of a sodium flame by diffraction with the image seen directly was 30°, the deviation should be something less than *10″*.

If you should wish me to try the new experiment, and would furnish the money (which I should estimate at most $500) I would be pleased to undertake it.

I should like to ask why, in a former letter, you said that the effect would be *compensated* if a *reflecting* grating used.

Very sincerely Yours
Albert A. Michelson

Michelson to Newcomb

Heidelberg July 8th 1881.

Dear Sir,

Yours of the 23d ultimate is just received and I proceed to answer your remarks in the order in which they appear.

1st. In measuring the length of the tape line, I used the comparator at the National Academy, bringing each time the middle of the division into coincidence with the cross hairs, whereby the error in each setting could not have been greater than 1/20 of the width of a division and was probably not more than 1/50, so that *even if the error was made in the same direction* through the whole tape the whole error would not be so great as 2/10000 as you have it.

The copy of the standard yard still remains at the Academy, so that you could compare it with your own, but it would not probably have anything like so great an error.

Possibly, if the temperatures at which the measurements were made were widely different, part of the error could thus be accounted for. As can be seen from the separate measurements of *one yard* in successive portions of the tape the values are quite different.

The error due to stretch was 0.00017, but this error did not occur in the comparison, as then only sufficient force was used to straighten the tape.

2nd. The error due to springing of the floor must be very small. If the spring were one inch the lengths could thereby be altered only 1/40000.

3rd. In measuring γ, the assistant was instructed to place a division of the tape over the center of a hole in the upper screw of the revolving mirror, and after every measure I went to the other end to verify its position; frequently, the arrangement was reversed I, myself, making this adjustment, the assistant reading off at the other end.

Each measurement was then corrected for half the thickness of the mirror.

4th. Not having my original work with me, I can not say what part of the tape was used as the zero, in the measurement of the distance. I believe it was the mark at 1 ft. *In none of the measurements was the brass ring used at all.*

5th. I am sorry to say that I fear notwithstanding instructions to the contrary, that the pier which supported the revolving mirror has been removed.

6th. If I can obtain the money and the time to repeat my own experiments or at least to get the apparatus in working order, I shall be most happy to do so.

Very sincerely Yours
Albert A. Michelson

Newcomb to Michelson

Aug. 15th [188]1

Dear Sir:

In reply to yours of July 2nd I thought I had already explained to you that when I said the effect of differences of wave lengths would be compensated in your experiment I meant to quantities of the first order. I did not suppose you would attempt so delicate a measurement of that of the second order.

I cannot furnish any money for the experiments as my appropriation is nearly exhausted. A decisive result with apparatus already in existence is so easily obtained that I can hardly doubt some one will soon attempt to reach it.

I have commenced getting flashes between Fort Whipple and the Washington monument, which, so far as yet reduced, seem to give the same result as the former series. I can find nothing wrong in the determinations of your apparatus. As soon however as the Naval Academy is re-opened I intend to learn whether your piers or their bases are still standing with a view of remeasuring your distance. The discordance you will perceive is still a mystery.

I would be glad to know if you intend to spend another winter in Germany especially as inquiries respecting your intentions were made at the Department the other day.

I have looked carefully through the points mentioned in your letter of July 8th it is not worth while to look at such small errors as those of the standard of comparison, correction for temperature, etc. But the measurement of γ and other matters connected with the revolving mirror are worth inquiry. You make no statement in your paper whether any of your adjustments were referred to the horizon, or the direction of the vertical. It seems you measured γ to the top of the upper screw of the mirror, which, judging from figure 5 page 119, was some five inches above the centre of the mirror. Did you make any determination of the angle between the two lines? Another allied question is whether the axis of rotation was parallel to the plane of the mirror and if not what the angle between them was. Supposing this angle to have been undetermined, the relation of the several axes and lines of reflection becomes a somewhat complicated one, having a material influence on the results.

As you describe your operations it would not appear that you aimed at any other adjustment of the relations of these various quantities than such as would cause the sun's rays reaching the micrometer by direct reflection to pass a fraction of a degree above or below the field of view.

In your original report your measures of distances are given in the form 100 ft minus different fractions of a foot, which fractions are never less than 0.1 and seldom greater than one foot. In other words, your line is measured in pieces most of which are 99 ft plus a fraction in length. This statement may assist your memory as to whether there is any possibility of a systematic error of 1/10 of a foot in either of your measures.

May I inquire where your revolving mirror now is? If you are going to remain longer in Europe I should like very much to determine the angle between its reflecting face and its axis of rotation.

Yours very truly,
Simon Newcomb

Newcomb to Michelson

Aug. 16th [188]1

Dear Sir:

Since writing you yesterday I notice that one of the points on which I touched needs more elucidation. On page 133 you have a = inclination of plane of rotation.

An inclination of one plane must be relative to some other plane but you give no hint as far as I can find to what plane you refer the inclination. Nor do you give any other definition of the term nor any statement how you determine it. I had always supposed that your inclination was simply the amount by which you had tipped the mirror to or from the observer in order to throw the direct image of the slit out of the field. But the value you assign, about a degree, is much greater than is necessary for this object. Moreover, on page 124 you say the revolving mirror was inclined to the right or left. But an inclination to the right or left around the axis passing through the line of sight would not prevent the reflection. You say further on that, when the image did not appear, the mirror was inclined backward or forward until it came in sight. This would of course change the angle of inclination and you say nothing about that change or how the final angle was determined.

I am sorry I could not have given your paper as critical an examination before it was finally prepared for the press as I have done since as I could have probably given you suggestions on a great many points respecting which questions would arise.

Yours very truly
Simon Newcomb

Michelson to Newcomb

Schluchsee, Schwarzwald Aug 29th 1881

Dear Sir,

Yours of the 15th and the 16th instants are at hand, and I proceed to answer in order. I intend to remain abroad unless something unforeseen should happen, until the first of next August, and shall spend the coming winter in Paris. My address will be, Care of Brown, Shipley & Co. London.

The adjustments were not referred to the horizon, but to the plane of the instrument, i.e., fixed mirror, revolving mirror, slit; but this plane differed very little from that of the horizon, I estimate not more than two or three tenths of a degree.

I admit that I completely overlooked the correction to be made in measuring to the top of the screw instead of to the centre of the mirror. If the distance is 5 inches (this would have to be diminished by something over half the length of the slit) the result would have to be diminished by 1/12000 or about 25 kilometers.

The axis of rotation was as nearly parallel to the axis of rotation as the instrument maker could make it. The plane of the mirror was flush with the face of the steel ring which enclosed it, and this was of course parallel with the axle.

The only remaining thing which could affect the parallelism would be the eccentricity of the pivot points. This error could not have been much greater than a tenth of a millimeter and if the length between the pivots was 100 mm the error would be less than 4 minutes. It appears to me, however that even if this error were as much as one degree the resulting inaccuracy would still be of the order of a ten thousandth.

I am reasonably certain that I could not have made a mistake of 1/10 of a foot in my measurement of the base line. The fact that the measurements were of the form 100 ft less a quantity greater than 0.1, makes it tolerably certain that the starting point was 0.1 on the tape.

I think that you will probably find either the base of the pier or at any rate the hole for the corner posts, and these would give from the drawing of the house, the position of the pier. If the machinist of the "Phlox," whose name I cannot now recall, or "Mike" the foreman of the yard is still there, they would be the persons who would be of most assistance in locating the spot.

The result which you obtained was, I believe, greater than 299700 and less than 299750. My result corrected for the error which you pointed out 299915, so that there is a difference of 200. Now the result of a systematic error of 0.1 in the use of tape, would represent an error of 300 in my result, so that our results would differ by 100 kilometers in the other direction.

The revolving mirror is in charge of Mr. A. G. Heminway,[41] No. 11, Wall St. New York, and you can obtain it by writing to him.

The inclinations, α, of the plane of rotation was referred to the plane of the instrument. This inclination was correctly stated to be "to the right or left," but it would have been more explicit to state that the plane in which the axis was inclined was at right angles to the line bisecting the [angle slit, revolving mirror, fixed mirror.][42] An inclination to or from the observer would not have the desired effect for it would throw the image of the slit as well as the direct reflection out of the field.

This movement is simply an adjustment which places the axis in the plane at right angles to the bisecting line.

[41] Michelson's father-in-law.
[42] Michelson's insertion.

The manner in which *a* was measured was as follows,—The revolving mirror reflects the light from the slit to the wall toward the fixed mirror. In this wall are two openings; H, where the light enters to go to the slit, M, where it goes to the fixed mirror.

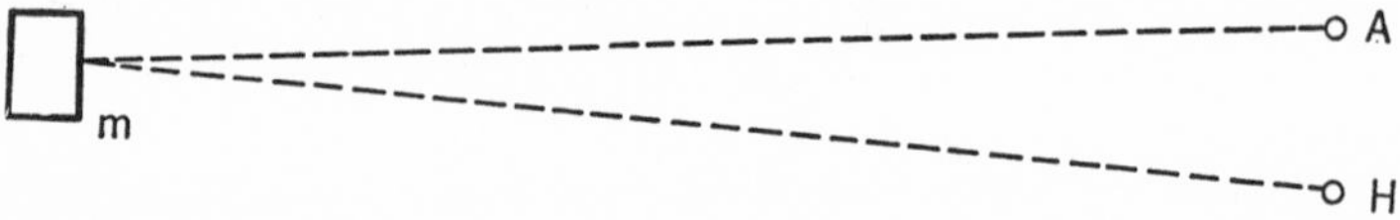

When the plane of the mirror is at right angles to the bisecting line, the image thrown on the wall exactly covers M. When the mirror is revolved till it is at right angles with the vertical plane passing through the mirror and slit, the image is above H, at A. The angle AMH which to be sure cannot be measured in this manner down to minutes, is the angle, *a*, required.

I freely admit that these points are far from being self evident, and cordially agree that your suggestions would have been of great value in the preparation of my paper.

Very sincerely yours
Albert A. Michelson

Michelson to Newcomb

2 Rue de Colisee Champs Elysees.
Paris, Oct 23d 1881

Dear Sir,

Your last was duly received. Would be obliged if after you have examined the instruments, you would send all of them to Mr. Heminway.

Have you had an opportunity of visit[ing] the U.S. N.A. to ascertain if the piers were still standing?

Would be obliged also if you would let me know the final results of your experiments as soon as they are completed.

I have been appointed a member of the Jury of the Int. Exhibition of electricity,[43] in which capacity I have had the pleasure of meeting quite a number of eminent men, as well as a fine opportunity of studying the recent advances in the application of electricity.

[43] The same international meeting where Rowland had the great success demonstrating his gratings described previously.

Hoping to hear from you soon, I remain

Very respectfully yours
Albert A. Michelson

Michelson to Newcomb

2 Rue de Colisee Paris March 22, 1882

My dear Sir,

Your letter of the 7th instant has been duly received. I do not think that too much importance should be attached to the loss of the revolving mirror, for it can be shown that the error you suggested must be very small. The price, of the mirror was I believe less than $40, so that would be the smallest consideration in repeating the experiments.

I cannot as yet promise to undertake the experiments, and if I were to do so, I should be obliged to make the conditions; first, that they be performed in Cleveland; second, that I should have at least $1000 at my disposal.

If, however, you desire that Capt. Sampson should undertake them, I shall be happy to give any suggestions that could be of value. The mirror was made by Fauth, and he could easily make another from the design in the published paper. The lens is in the "Cabinet" at Annapolis.

I intend returning to New York in about two weeks, and shall in all probability pay you a visit, when we can discuss the question at leisure.

Very truly yours
Albert A. Michelson

Michelson to Newcomb

New York June 8th 1882

My dear Sir,

Your letter of the 7th instant has been duly received. The sum of $1200 which you have applied for is in excess of my estimate, which, if my memory serves, was only $1000. But if you have $1200 *for the purpose,* I think I could guarantee the successful establishment of the apparatus.

The advantages in making the experiments in Washington are I believe; 1st saving of the cost of an engine and blower; 2nd benefit of your supervision.

A separate building would be needed in any case. As for assistance, that I could also have without cost in Cleveland, or at any rate, the cost would be very trifling.

It will be impossible for me to go now to Washington, but even if I could do so, as I would be there but a month I could not promise even to have things in working order.

I think that within two months I could have the apparatus in working condition in Cleveland and in three months to have a satisfactory result. At the end of that time there would probably be enough money left to pay the expenses of a visit should you desire to make one in order to make measurements yourself. I think it quite possible to borrow an engine, or at least to rent one for one or two months. If you undertook the experiments or had someone else to do so, I hardly think you could complete the experimen[ts] in less time.

It will probably be somewhat inconvenient for me to carry out the experiments,[44] bu[t] I am willing to undertake them, partly because I should be able to try the experiment with liquid.

Should you still think it advisable to try them at Washington as I said before, I would be happy to render, by writing, such assistance as I could.

very respectfully Yours
Albert A. Michelson

Michelson to Newcomb

New York June 12th [18]82

Dear Sir,

Yours of the 10th just received.

I. The distant reflector used in my former experiments was the heliostat mirror used for Transit of Venus photographs. Could not the same apparatus be used again? If not the mirror which you had made would answer the purpose.

In measurement of the speed of mirror I employed a fork bearing a steel mirror which was kept vibrating by an electromagnet of pitch Vt_2 and a standard Vt_3 fork. These were made by Koenig of Paris. In place of the standard Vt_3 a standard Vt_2 would be preferable. If these could be borrowed from the Naval Academy or from the University of Pennsylvania it would save the trouble of sending to Koenig.

It would also be necessary to have a small heliostat similar to

[44] To confirm whether light is wave-like or corpuscular.

that you employed. If any apparatus arrives in time from Paris, I may be able to get the use of the one which I ordered.

A bellows which I used for regulating the blast of air is I believe still at the Naval Acad. but I think I will not require it.

II. The revolving mirror is, I think, about as large as necessary, but I think it would be advantageous to but make the disc *C*, (Fig 5—of my paper) about twice as thick as represented.

If the additional cost would not be too great, it might also be advisable to have a box and turbine similar to TB, Fig 4, placed on the lower part of the axle, (the latter being elongated for the purpose) in order to give a rotation in the opposite direction.

It would be convenient, too, if the frame were so arranged that the whole of the rotating part could be more easily removed than was the case in the old form. The thickness of the ring R, had better be increased by mm.

III. The diameter of the distant mirror may be anyth[ing] over 8 inches, of course the larger the better. There would be an advantage in having it curved to a radius of 2000 feet, as I think that will be the best distance to use.

The mirror (provided you cannot obtain the one previously used) may be mounted in the same manner as your own, except that I should like to have a sextant-telescope attached to the frame which holds the mirror, by a universal joint.

IV. The only suggestion I would have to make is that the apparatus may be made by Alvan Clark & Co. and that they be given all the liberty they desire in the execution.

Am I to understand that I am authorized to put up the building, hire or purchase the engine, etc, on my arrival in Cleveland, or am I to await further instructions?

Very truly yours
Albert A. Michelson

Michelson to Newcomb

New York June 15th 1882

Dear Sir,

In answer to yours of the 14th, I start for Cleveland on July 1st.

The mystery is solved and the revolving mirror found:

I mentioned at the dinner-table that I was to repeat my experiments, and that naturally led to the subject of the lost instrument.

Mrs. Heminway asked what I intended to do with the other

instrument. "What other?" I asked. "The one down stairs." I knew immediately that the mirror was found. Mrs. H. knew that I had two instruments in New Rochelle, and *did not* know that I had left one of them (the micrometer) at the Steven's Institute, and has been thinking (not knowing a "revolving mirror" from a great equatorial) that *that* was the missing instrument.

I measured the angle of the plane of the mirror with the axis and found it less than two minutes.

If it is desirable to have rotation in either direction, I think it would be advisable to construct a new instrument; for it would cost almost as much to alter the present one for the purpose.

The present one, if the double rotation is not required, would have to be overhauled, for the pivots and sockets are somewhat worn.

I think if the Clarks were *willing* to work on it and would not take too long, the job would be very much more satisfactory than if done by Fauth.

What has become of the 9 inch mirror which you had made three years ago?

Very respectfully yours
Albert A. Michelson

Michelson to Newcomb

Cleveland Sept 6 [18]82

Dear Sir,

I hope you will excuse my rather long silence in view of the fact that I was anxious that my next letter should give an approximate result of my experiments.

I have succeeded in making some fifty observations of extremely varying degrees of accuracy and the result thus far is 299810 (vacuo).

I think it quite possible that the final result may vary from this as much as 100 kilometers, but not much more.

My arrangements are as yet far from perfect and unfavorable skies have prevented my doing much work.

I have attempted to suppress the bellows which I used in my previous work as a regulator but find now that they will be necessary.

The mirror turns now so smoothly that it is with difficulty that any sound can be heard even when the ear is but two or three inches away.

The atmospheric conditions are not so good here as they were at Annapolis, and I have thought it might be advisable to use the electric light. If I can, will try to persuade some of the Brush people to lend me a machine but if that cannot be done could you afford to rent one? The rent is $200 per year and they do not rent for a shorter time.

Could you give me an idea of how much we have spent thus far?

When do you expect to retu[rn] from your trip, and will you pay a visit to Clevelan[d]?

Have on hand but $20 of the fund you sent, can you let me have more before you go?

I have discovered a mistake in my previous work of 12 Kilometers, *P* 141, "measurement of ϕ." The correction +12 Km. should be cancelled. This together with the correction you pointed out of 25 Km. would make my true result *299910.*

Yours very truly
A. A. Michelson

Michelson to Newcomb

Cleveland, O. Sept 12 1882

Dear Sir,

The conditions have been unfavorable for observations, since my last writing, but I have succeeded in getting about twenty fair readings, and the result of these is 299860. There is a slight discrepancy between my tape and a surveyors tape 300 ft. long which I have been using.

If I take my own tape as standard the result is about 20 kilometers greater.

Have not detected any "slipping seconds" in the clock.

Have succeeded in borrowing a "Brush machine" and will try to work with that in cloudy weather. My school duties commence on the 20th, so that I may not be able to complete the experiments as soon as I thought I could.

Money has given out.

Very truly Yours
Albert A. Michelson

Michelson to Newcomb

Cleveland, O. Sept 17 1883

Dear Sir,

The report of my redetermination of the velocity of light is ready and I will send it by mail tomorrow.

The repetition of Foucault's experiment is also completed, and as you will see by the report, the result not only confirms Foucault's qualitative result, but agrees very closely with theory which requires that the ratio of the two velocities should be equal to the index of refraction of water.

I have not yet been able to make the experiment with carbon disulphide (for different colors) and would like to know whether I can make use of part of the fund remaining, for that purpose.

I have resold the engine, as per agreement, and have remaining about $220.

Very truly yours
Albert A. Michelson[45]

Nautical Almanac Records, National Archives

Michelson and Morley

Michelson's Potsdam experiment on the motion of the ether relative to the earth did not satisfy many scientists. To settle the important issues raised by his negative results, he was encouraged to redo the experiment with an even greater degree of precision. With the aid of a grant from the Bache Fund, probably obtained through Newcomb, Michelson and Edward Williams Morley (1838–1923) of Western Reserve College performed one of the great experiments of physics in 1886–87.

Even before this, Michelson had joined forces with Morley in careful tests to determine whether light had a corpuscular or a wave constitution. Morley was a chemist with no significant formal training in science. A graduate of Williams College and Andover Theological Seminary, his first ambition was the ministry. But his qualifications, no worse than those of many other American college teachers, were satisfactory to a struggling Midwestern school. Lack of equipment and of a suitable professional milieu, as well as lack of formal training, hindered his scientific career. But his collaboration with Michelson certainly demonstrated a considerable talent as an experimentalist.

After the Michelson-Morley experiment of 1886–87 confirmed the earlier negative result, the two continued to work together until 1889, when Michelson left Case Institute to go to Clark University. Morley then resumed in earnest his great project—the determination of the relative weights of oxygen and hydrogen. Facilitating his investigations was a precision balance that Michelson obtained for him from the Smithsonian Institution. Like the work of T. W. Richards (1868–1928) at Harvard,

[45] P.S. omitted.

the first American to win the Nobel Prize in chemistry, Morley's research was notable for precision and experimental niceties. Neither Morley nor Richards developed the theoretical implications of their findings—that the elements were composed of isotopes of differing weights. That task was done by Europeans.

Michelson, the first American to win the Nobel Prize in physics, was also a master experimentalist. The consequences of the negative result of the Michelson-Morley experiment were worked out in Europe. These culminated in Einstein's work. Many observers on both sides of the ocean were inclined to characterize Americans as lacking in generalizing capabilities and to ascribe their skill in experimentation as an extension of the presumed national facility with mechanical devices. This explained neither the small number of American experimental physicists nor the appearance of a J. Willard Gibbs.

Michelson to J. Willard Gibbs

Cleveland, O. Dec 15 1884

My dear Professor Gibbs,

I have delayed so long the acknowledgement of your kindness in sending me your papers on "Electromagnetic theory" and "Vector analysis," that I am a little in doubt whether it is still "better late than never."

I have had such a press of work that thus far I have not had leisure to study them but hope to do so very soon.

I had the pleasure of talking with Sir Wm. Thomson and Ld. Rayleigh,[46] on the subject of influence of motion on media on propagation of light, and they both seem to think that my first step should be to repeat Fizeau's experiment.[47]

I have accordingly ordered the required apparatus and hope to be able to settle the question within a few months.

Meanwhile, I should very much like to have an opinion from you concerning the feasibility of the experiment which I described to you rather hastily at our last meeting, Namely: 1st Granting that the effect of the atmosphere may be neglected, and supposing that the earth is moving relatively to the ether at about 20 miles per second would there be a difference of about one hundred-millionth in the time required for light to return to its starting point, when the direction is parallel to earth's velocity, and that, when the direction is at right angles? 2nd. Would this necessitate a move-

[46] William Thomson, Lord Kelvin (1824–1907), the most prolific English physicist. Lord Rayleigh (1842–1919) was a great English physicist.

[47] Armand Hyppolyte Louis Fizeau (1819–96), an experimental physicist. The experiment, done with Morley, involved sending light through water.

ment of the interference fringes produced by the two rays? 3rd. If these are answered in the affirmative, provided the experiment is made so far above the surface of the earth that solid matter does not intervene, what would be the result if the experiment were made in a room?

Trusting that I am not imposing on your good nature by proposing these conundrums and hoping to hear from you, I remain

Very sincerely yours
Albert A. Michelson

Gibbs Papers, Yale University

Edward W. Morley to Sardis Brewster Morley[48]

Cleveland, Ohio April 8, 1885.

Dear Father:

Your letter with the draft came this morning. A couple of weeks ago, I took the first cold I have had this winter, at least the first one which amounted to anything. This has been a severe one, and it is now at the stage when I do not sleep well at night, on account of the irritation in my throat, making me cough. The cold is near the end, however.

Rev. George R. L. Leavitt, who was at Cambridge, was settled here last week, over the church where Charles Terry Collins was pastor. I went down to the examination. There is a reception to him and his wife this evening, but my cold kept me from going. I am sure Leavitt will be the leading Congregationalist minister here, if he has his health, and stays. There is a debt on the new church, and I suppose Leavitt will want to get that paid the first thing.

Something will have to be done to our church this summer. It is rather dingy inside, and there is not enough light. Some want to tear down, and build a new church, but this is not twenty years old. Some fifteen thousand dollars has been offered towards building anew; but this is far from enough, and the men who offer it want to get the Society committed to re-building, with the idea that then those who do not want to re-build will have to take hold and pay the chief part of the cost. I do not know what plan will be adopted. It would not be a good plan to load ourselves with

[48] In the subsequent letters of Morley to his father only the sections on his work are given. The remainder of the letters are on family matters, church affairs, and the like.

a debt, for there are so many new churches near us, that people would begin to go away from us to avoid helping pay their share of an expense they did not approve.

Our vacation ends tomorrow morning. A week's intermission has been very acceptable.

I think that I have written that Mr. Michelson has been getting ready an apparatus for an experiment at our building. He did not calculate very well and will have to get alterations made before he can do anything. So it is likely to be a week before he can make his experiment. His lack of calculation was rather curious. It would have been so obvious, if he had only stopped to think.

Our girl is sick, so that Belle has no help for a few days. The girl may come back tomorrow. I guess not much is the matter. Belle sends love to Lizzie and you in which I join.

Your affectionate son,
Edw. W. Morley.

E. W. Morley to S. B. Morley

Cleveland Sept. 27, 1885

. . . Mr. Michelson of the Case School left a week ago yesterday. He shows some symptoms which point to softening of the brain; he goes for a year's rest, but it is very doubtful whether he will ever be able to do any more work. He had begun some experiments in my laboratory, which he asked me to finish, and which I consented to carry on. He had money given him to make the experiments, and he gave me this money: the experiments are to see whether the motion of the medium effects the velocity of light moveing in it. I made some trials last week, and found that a good deal of modification in the apparatus would be necessary, and made some drawings for the apparatus. As soon as these are done, I shall try again. When every thing is ready, it may not take more than a week to make the experiment . . .

Morley Papers, Library of Congress

Michelson to Newcomb

Hotel Normandie New York Sept 28, '85

My dear Sir,

I have been obliged by ill health to give up all work for at least a year, and have turned over the money remaining from the

Bache Fund to Prof. E. W. Morley of Cleveland, who is quite competent to finish the experiments should I be unable to do so.

The sum was $75 to which I have added $100 of my own.

Hoping that in case of your coming to New York I may have the honor of a visit, I remain

Very truly yours
Albert A. Michelson

Newcomb Papers, Library of Congress

E. W. Morley to S. B. Morley

October 12, 1885

. . . I have nearly got ready for the experiment which I am to make for Dr. Michelson. I lack a pump, in place of one which he got, and which was not suited to the requirements. A man has gone to Chicago, who will send me a pump this week, if nothing prevents. Then I shall be ready to make the experiment. Whether it will take a day, or a week, or a month, no one can tell beforehand. So far, I have not done much this year, but have taken things pretty easily. Soon I shall get to work on the air analysis, I presume . . .

Michelson to Morley

Hotel Normandie New York Oct 12 [18]85

My dear Prof. Morley,

I am here under the care of Dr. Hamilton, who assures me that my trouble is not serious, and that he hopes that in a month or two I shall be recovered.

My treatment consists principally in amusing myself in whatever way I like, which, by the way, is not so trying as one might imagine, when one gets used to it. And in a course of "massage" twice a day. The latter is not quite so enjoyable but I shall become accustomed to it in a short time.

Have you had time yet to experiment on the "fringes"?

Let me know how everything is going on and how the two institutions agree; in short any thing and every thing that may interest us both.

With kind regards to all mutual friends

Sincerely yours
Albert Michelson

Michelson to Morley

Hotel Normandie N.Y. October 23d 1885

My dear Prof. Morley,

Your welcome letter arrived this morning. I am a little disappointed to learn that some one had been found to take my place, as I had just written to Mr. Abbey that the doctor thinks I could have resumed my duties in a few weeks.

Am very much pleased to know that your work with the "fringes" has been so successful, and I think you have done wonderfully well to get along so far, and am not in the least impatient.

In selecting the glass for the brass tube, I think it will be sufficient, if the piece chosen does not distort the fringes, tho' perhaps it would be well to examine a distant object by reflection from its surfaces with a telescope.

I should like to hear everything you have time to communicate as I am so much better that it will not tire me in the least.

With kind regards to all

Very sincerely yours
Albert A. Michelson

Morley Papers, Library of Congress

Michelson to Rowland

Hotel Normandie, N.Y. Nov 6th [18]85

My dear Professor Rowland,

Your very welcome letter duly received, many thanks for your kind wishes.

The trustees of the Case School have, I am sorry to say, added another to the mistakes they have made, in appointing to fill my place (temporarily, however) a man who has had "twenty years' experience as a teacher" but who has no particular specialty.

Now that my physician assures them that I will be able to resume work in a few weeks they wish to have me return; but in as much as their man has been appointed for a year they wish me to "take into consideration the strained condition of their finances," which strained condition results from the very action they were forewarned against, namely, "putting their money into bricks instead of brains."

It looks to me as tho this condition will last indefinitely, and taking into consideration the inconvenience of being so far removed from scientific centers, I would much prefer a position farther East. Please to give me your advice in this matter, and if anything fitting my capabilities should turn up elsewhere be kind enough to let me know.

Very sincerely yours
Albert A. Michelson

Rowland Papers, Johns Hopkins University

E. W. Morley to S. B. Morley

November 19, 1885

. . . When Professor Michelson was taken sick and forced to go away, he asked [me] to complete an experiment which he had begun. Things have gone slowly, and not much is done yet: Last week we filled 30 carboys with distilled water, which may not be quite enough. This week I hope to get a pump set up with which to pump this water into a tank in the attic: if the pump works well, I may get the experiment done in a week. One pump failed, and we had to get another. Mr. Michelson is better, and means to return in December: but may not want to do any work other than his teaching . . .

E. W. Morley to S. B. Morley

January 31, 1886

. . . Mr. Michelson and I get on somewhat with our experiment. Last week we took apart our apparatus and lengthened the tubes in which the water flows. Since that we have not had good success: our light has not been sufficient to see very well through so great a length. On Saturday, I told the janitor to distill us some thirty carboys of water: with this clean water, I presume we shall succeed well. It will take a week to get this water. After that we ought to get through in a week . . .

E. W. Morley to S. B. Morley

April 17, 1887

. . . Michelson and I have begun a new experiment. It is to see if light travels with the same velocity in all directions. We have not got the apparatus done yet, and shall not be likely to get it done for a month or two. Then we shall have to make observations for a few minutes every month for a year. We have a stone on which the optical parts of the apparatus are to be fixed. This stone is five feet square, and about fourteen inches thick. This we shall have to support so that it can be turned around and used in different positions, and yet it must not be strained differently in the different positions. Now since a strain of half a pound would make our observations different useless, we have to support it so that its axis of rotation is rigorously horizontal. My way to secure this was to float the stone on mercury. This we accomplish by having an annular trough full of mercury, with an annular float in it, on which the stone is placed. A pivot in the centre makes the float keep concentric with the trough. In this way I have no doubt we shall get decisive results . . .

E. W. Morley to S. B. Morley

June 2 1889

. . . Professor Michelson has resigned at the Case School, and is going to the new Clark University[49] at Worcester, Massachusetts. As I have worked with him a good deal, and am now right in the middle of a long piece of work with him, this is a good deal of a disappointment to me. The teaching at Worcester will be much more to his mind than the teaching here, and they are glad to have him go from the Case School. In fact all the professors except in chemistry will leave at the end of this year. What it all means, I hardly understand. They certainly lose one of the first two physicists in the country in losing Michelson. But it may be that he is not a good teacher for the class of students whom they get there now. . . .

My work on the oxygen is getting along somewhat now. I have at last got what I have tried to get since last autumn, namely, pure hydrogen, which is the first . . . thing in getting the atomic weight of oxygen. But it is so near commencement, that I shall not have time to accomplish much, I fear. . . .

[49] G. Stanley Hall (1846–1924) the psychologist, whose success at Johns Hopkins helped block C. S. Peirce's academic career, was the head of Clark. He aimed to surpass Hopkins but did not succeed. Michelson went to the University of Chicago in 1894.

E. W. Morley to S. B. Morley

June 9 1889

. . . I had a letter on Friday which offered me the professorship of general chemistry in the University of Michigan at Ann Arbor. The salary is twenty-five hundred dollars. Here I now get three thousand, but it is likely that they will cut all salaries of that amount down to twenty-five hundred; so that I should be just as well off pecuniarily at Ann Arbor as here. The teaching there would be more to my mind than it is here, since it would all be in the line at which I am best. I am going there next Thursday or Friday to talk it over, and prepare to make up my mind. How I shall decide, depends, of course, on a good many things which I do not now know, but I am strongly inclined to go. I have been here twenty years; if I do not go now, I shall not have another opportunity, probably, and I am not quite ready to make up my mind to stay here till I can teach no longer. Dr. Michelson, who is going away this summer to Clark University at Worcester, tells me to go by all means, and tells me that I am not appreciated here; which may not mean much. But it would be pleasant to go to a place where I was not the only man with the impulse to occupy myself with research. . . .

Michelson to Morley

June, 29 1889

My dear Professor Morley

I am delighted to know that you were able to make such good terms, and under *these* circumstances think you did well to stay.

I got another letter from Rowland in which our results are *confirmed* with the exception of the yellow, which is noted as "indistinct." The numbers agree to about one two-hundred-thousandth, or less (I haven't the figures here).

Have begun making requisitions for apparatus, and plans for laboratory etc. There is much to weary one, but that must be so in every new establishment, and I think within a year I'll be moderately comfortable.

Would it be possible for you to send me a copy of Marquarts catalogue?

Yours cordially
Michelson

Morley Papers, Library of Congress

J. Willard Gibbs

The two Drapers came from the old tradition of the medical school acting as a poor substitute for the graduate school. The engineering school with its mixture of mathematics and practical vocationalism produced Rowland. Newcomb's astronomy and that of many of his contemporaries was a late harvest of the fruits of the geophysical tradition. Michelson, the Navy scientist, was the greatest product of that disappearing scientific milieu. Morley, the minister who became interested in natural history and then drifted into chemistry, was very much like the amateur botanizing, geologicizing clergyman of an earlier day in his origins. Josiah Willard Gibbs (1839–1903) was a unique product of one of the few academic scientific traditions in the United States.

His father, of the same name, was a professor of philology at Yale, and the son spent his life on the New Haven campus. Not that Yale was a particularly suitable environment for the birth and development of genius in mathematical physics. On the contrary, every factor supposedly adverse to the development of theoretical sciences in America was present. But at Yale an academic tradition existed encompassing both classics and science. By chance or design, it was possible for an abstruse theorist to find a haven.

The tradition dated back to the elder Silliman and was maintained after his death by his son-in-law, James Dwight Dana. The two were largely responsible for the establishment in 1846 of the first successful graduate school in the United States, the Department of Philosophy and the Arts. An undergraduate scientific school was founded in 1854 and renamed the Sheffield Scientific School in 1861. That same year the graduate program was expanded to include studies for the doctorate. Much of this development and expansion took place while Gibbs was a student at Yale.

The Sillimans, Dana, and their associates entertained views on the relations of pure and applied research typical of the day. They were equally enthusiastic about both and often confused them. Dana was interested in providing a scientific education for collegians to replace the old classic curriculum. He had no aversion to applied science and apparently looked with equanimity upon the growth of agricultural chemistry and engineering in the Yale complex. Between a growing engineering tradition and an unyielding humanities tradition, the pure sciences tended to lose ground to other fields in Yale in the half century following the inauguration of doctoral training. At the same time other universities were expanding the sciences vigorously. None of its rivals, however, could boast an equal to Gibbs.

A prophet predicting the arrival of a scientific genius in New Haven shortly before the Civil War might have reasonably guessed the field as geology, applied chemistry or engineering— not theoretical physics. Gibbs actually did his doctoral dissertation in 1863 in engineering—"On the form of Teeth of Wheels in Spur Gearing." He later patented a braking device for railroad trains and invented a steam-engine governor. Personal inclination and the influence of teachers brought him to objects more suited to his talents. Of these teachers the most important is Hubert Anson Newton (1830–96), a mathematician noted for his studies of comets and meteors. Gibbs and Newton were lifelong friends. Another teacher who possibly influenced Gibbs was Elias Loomis, the old astronomer-meteorologist-mathematician now teaching at his alma mater.

From 1866 to 1869 Gibbs studied in France and Germany. In 1871 he was made a Professor of Mathematical Physics in the Department of Philosophy and the Arts at no salary (like Marsh, the great paleontologist). Gibbs had produced nothing in mathematical physics at that date. Depending on one's frame of reference, the action of the university can be described as highly perceptive or a lucky gamble. Gibbs proceeded to produce a stream of monumental theoretical works, some appearing in such an unlikely location as the *Transactions* of the Connecticut Academy of Arts and Sciences where they were not immediately noted by his peers. But Gibbs went to great pains to send reprints and even page proofs to leading scientists. Among the first to appreciate Gibbs and to spread word of his findings was Maxwell in Cambridge University.

Recognition by scientists at home and abroad did not change conditions at Yale. Gibbs was still unpaid and had few students. In 1879, as detailed in the letters that follow, Rowland recommended Gibbs to President Gilman for a post at Johns Hopkins. Gibbs was inclined to accept at first. Only Newton knew of this and, presumably, he interceded with influential individuals at Yale to get a counter-offer. Gibbs remained in New Haven.[50] Johns Hopkins was a livelier place then; contacts with Rowland, Remsen, and Sylvester could very well have stimulated all concerned; more first-rate students would have profited from his instruction.

Gibbs, of course, is the great exception to any generalization about American disinterest in or incapacity for theoretical work in science. About him certain legends have gathered, often concerning his personality and his supposed lack of recognition. Shy, impractical, and meek are words used or implied in descriptions of Gibbs. In fact he had warm friendships and was reasonably affable in society; his letters at Yale on the management of the family funds are far from impractical; and the firmness with which Gibbs acted in matters important to him, that is, as a scientist, is far from meek.

During his lifetime, Gibbs sought and received the recognition of his scientific peers, both in America and abroad. This is what was important to him and why Gibbs firmly upheld his views and rights in scientific disputes. To a man doing theoretical work in thermodynamics and sta-

[50] He had received his first pay for teaching science from the university that year for taking over Newton's classes. Gibbs had a modest private income.

tistical mechanics, developing vector analysis, and, in general, laying the foundations for physical chemistry, it was possible literally to communicate his work only to few scientists. During his lifetime most of these were Europeans; few Americans were both interested and competent to follow his work. The first consequences, theoretical and applied, of Gibbs' researches occurred in Europe and were later assimilated by a maturing body of American physicists and chemists. In the early decades of the nineteenth century, the elder Silliman could talk to the public and acquire popular recognition; by the last decades Gibbs could not truly communicate with the public; his recognition could come only from a small group of fellow specialists.

Rowland to Gilman

Baltimore May 8th/1879

Dear Sir,

Some time since I called your attention to the necessity of having the subjects of Mechanics and Mechanical Drawing taught in the University. I would like at this present time to call your attention to Prof. Willard Gibbs of Yale College as an excellent candidate for the chair of Mechanics not only from his eminence but because he will fill a wide gap in the mathematical staff. There are two methods of considering such a subject as mechanics, the one being purely mathematical and the other a combination of the mathematical and physical, and they must be cultivated by minds of different kinds. Of such men I may mention Cayley, and Sylvester[51] on the mathematical side and Helmholtz, Thomson and Maxwell on the mathematical physical side and in the latter class should also be included Newton who was indeed the beau ideal of this class of minds. Prof. Willard Gibbs belongs preeminently in the latter class; to men who not only grasp the subject from a mathematical standpoint but who see the subject in *all* its bearings, and to whom the problems of nature are something more than targets on which to practice with their mathematics.

As to his eminence there can be no doubt. Maxwell, in his small work on the "Theory Heat" has introduced no less than thirteen pages from the papers of Prof. Gibbs, and he told me personally that the new methods of Prof. Gibbs allow problems to be solved which he, Maxwell, had almost concluded were incapable of solution.

In conclusion I may add that Maxwell, who has the best oppor-

[51] James Joseph Sylvester (1814–97), the English mathematician who came to Johns Hopkins. Arthur Cayley (1821–95) was his friend and collaborator.

tunity for judging in the world, states that this class of men is very rare while brilliant men of the purely mathematical class are comparatively common.

Yours Respectfully
H. A. Rowland

Gilman Papers, Johns Hopkins University

Dana to Gibbs

New Haven April 26 1880

My dear Prof. Gibbs:

I have only just now learned that there is danger of your leaving us. Your departure would be a very bad move for Yale. I have felt, of late, great anxiety for our University (using a name we are striving to deserve) because there seemed to be so little appreciation among our Graduates as to what we need and so few benefactions in our favor; and now the idea of losing the leading man in one of our departments is really disheartening. I do not wonder that Johns Hopkins wants your name and services, or that you feel inclined to consider favorably their proposition, for nothing has been done toward endowing your professorship, and there are not here the means or signs of progress which tend to incite Courage in Professors and multiply earnest students. But I hope nevertheless that you will stand by us, and that something will speedily be done by way of endowment to show you that your services are really valued.

Johns Hopkins can get on vastly better without you than we can. *We can not.*

Sincerely yours
James D. Dana

Franklin B. Dexter to Gibbs

Yale College New Haven, Connecticut
28. Apr., 1880

Dear Sir,

I am directed by the Prudential Committee of the Corporation to express to you their strong desire that your connection with the College will be continued, and to say that they have voted to recommend to the Corporation at their next meeting that after the current year an annual salary be attached to the chair of

Mathematical Physics, of not less than two thousand dollars. It is their hope that the salary will be fixed for next year at twenty-five hundred dollars, and that immediate steps will be taken to secure an endowment of the chair which will yield a full salary.

I may also add that the Academical Faculty have in contemplation a proposition that you should be invited to offer an optional course in their department; and that they, as well as all the other gentlemen connected with the Philosophical Faculty, would regard your removal as an irreparable loss.

Very truly yours,
Franklin B. Dexter,
Secretary. [of Yale]

Gibbs to Gilman

New Haven Apr. 29 1880

My dear Sir

Within the last few days a very unexpected opposition to my departure has been manifested among my colleagues, an opposition so strong as to render it impossible for me to entertain longer the proposition wh[ich] you have made. I had not previously spoken of the matter outside of my own family except in confidence to Prof. Newton, as he was leaving for Europe. I only mention this to explain why after so long a time I have arrived at a decision contrary to that to wh[ich] I was tending.

I remember your saying that you told Prof. Sylvester that you thought it would be hard for me to break the ties wh[ich] connect me with this place. Well, I have found it harder than I had expected. But I cannot omit to say that I am very sensible of the cordial sentiments wh[ich] have been expressed by yourself as well as by other members of your University; I mean especially Professors Sylvester and Rowland.

Very truly Yours
J.W.G.

J. J. Thomson[52] to Gibbs

Trinity College Cambridge July 13th 1884

Dear Sir,

I must apologize for not writing to thank you for your paper on the "Equilibrium of Heterogeneous Substances" before, but I

[52] Sir Joseph James Thomson (1857–1940), an English physicist.

have only just returned to Cambridge after a holiday. I am very much indebted to you for the paper which I am now busily engaged in reading. I am especially interested in it as I have lately been working at a theory of chemical combination: though my method is quite different from yours in the only case I have compared my theory with yours viz the effect of the relative quantities of the combining substances on the Chemical Equilibrium. The numerical results are almost identical. I have not got to the theory of Electrolytic Equilibrium in your paper yet, but I can see that it will be useful for me for my report which I am sorry to say I cannot get finished in time for the Montreal Meeting of the B. A. Believe me,

Yours very truly
J. J. Thomson

Asaph Hall[53] to Gibbs

1886 Sept. 2nd

Professor J. W. Gibbs
Yale College

I have just read the notice in "Science" of your address at Buffalo. You need have no fear, I think, that astronomers will not adopt any real improvements in their method of computing. In fact it is curious to see how a set of computers will find out the easiest way to do a piece of work. This way is not always the shortest, but it's the one requiring the smallest amount of physical and mental labor. The computer is a real mercenary, and does not care for the reputation of anybody. He is like a stream finding its way down a mountain side, a good illustration of the principle of least action.

Can you not give us a paper for the autumn meeting of the Nat. Academy? Nov. 9th in Boston. If you could give us a memoir for printing so much the better.

Yours truly
A. Hall

Gibbs to Wilhelm Ostwald[54]

New Haven Oct. 26, 1888

My dear Sir

I hear through Mr. Loch [?] (now I believe in Newport RI)

[53] Hall (1829–1907), of the Naval Observatory, won fame as the discoverer of the moons of Mars.

[54] Ostwåld (1853–1932) was a German physical chemist and Nobel Laureate.

that you are desirous of obtaining a copy of my "Equilibrium of Heterogeneous Substances." My extra copies having been long since exhausted, it can only be obtained by purchasing Vol. III of the 'Transactions of the Connecticut Academy' of which it constitutes a large part (825 pp). This will be sent to any address by the secretary Mr. Addison Van Name[55] (New Haven Connecticut) on receipt of the price $6.00 by International Post Office money order or other-wise.

I send by book post a few minor papers on kindred subjects, of which I beg your kind acceptance.

I remain yours very truly
J. Willard Gibbs

Gibbs to Ostwald

New Haven Dec 7 1888

My dear Sir

I should be very glad to have my essays in Thermodynamics made accessible to a larger circle of readers. Yet I should have feared that the call for a German edition would hardly justify the labor and expense of the translator and publisher. If, however, you think differently, I should be glad to hear from you more definitely in regard to what you think practicable.

With thanks [to] your kind interest in my work, I remain,

Yours truly
J. W. Gibbs

Oliver Heaviside[56] to Gibbs

Paignton Devon April 6 1894

Dear Professor Gibbs

I enclose a letter I have had from Alex MacAulay, as he seemed to wish it though there is nothing particular in it. He seems to me to be a very clever fellow, and he knows it, and shows that he knows it a little too much sometimes. Or, maybe it is only his protest against neglect and discouragement he may have met with.

[55] Gibbs' brother-in-law, Librarian of Yale.

[56] Oliver Heaviside (1850–1925), British applied mathematician and champion of Gibbs' vector analysis against the views of the followers of Sir William Hamilton's quaternions. This struggle, referred to in Gibbs' Buffalo address raged fiercely in Britain. Knott and Tait are eminent British mathematicians on the other side.

Of course my primary object in writing the letter which has pleased him was not to please him in particular, but *to point* out pointedly by an example Prof. Tait's ignorance of the Vector Analysis he and his satellite have so contemptuously condemned. No one would imagine from Prof. Knott's abuse of your Vector Analysis that you had carried the theory of linear operators far beyond anything in Prof. Tait's Treatise.

On the general question, I should not be surprised to find that there will be some quaternionic activity and diffusion in the next few years, on account of the false idea prevalent that V. A. is (or ought to be) quaternionics. I say the reverse is true; that quaternionics is a spurious V. An. As, however, the associative property is attractive, it deludes, and quaternionics may, as I said, have some vogue. It is curious to think that a simple matter like V. A. should be, in a sense, more advanced than the highly developed quaternionics that men should go through the latter to get to the former!

Yours very sincerely
Oliver Heaviside

Gibbs Papers, Yale University

THE HIGHEST AIM OF A PHYSICIST BY HENRY A. ROWLAND [57]

This is Rowland's valedictory to science in America as first President of the American Physical Society. When he entered the field, a special organization for physics was almost unthinkable in America. At the end of the century physics was securely, if modestly, launched as a separate scientific field in America. His example and teaching contributed greatly to the development.

Now facing the certainty of death from diabetes (insulin was not yet known), Rowland stated his beliefs in blunt terms. Although there is ample recognition of the benefits ultimately derived from basic science, Rowland's address was an unyielding assertion that satisfying human curiosity was a justifiable end in itself. Flattering his audience, Rowland characterized them as an elite, an aristocracy of intellect. His points probably convinced no one but the already converted.

Speaking as a classical physicist, Rowland was seemingly unaware of the intellectual revolution about to convulse his field. Speaking as a good product of the nineteenth century, Rowland was most optimistic about the future. A great age of science had passed; a greater one was about to begin.

Gentlemen and Fellow Physicists of America:

We meet to-day on an occasion which marks an epoch in the history of physics in America; may the future show that it also marks an epoch in the history of the science which this society is organized to cultivate! For we meet here in the interest of a science above all sciences, which deals with the foundations of the Universe, with the constitution of matter from which everything in the Universe is made, and with the ether of space by which alone the

[57] Presidential address delivered at second meeting of the American Physical Society, October 28, 1899. Excerpted from the *Bulletin* of the American Physical Society, vol. 1, no. 1.

various portions of matter forming the Universe affect each other even at such distances as we may never expect to traverse whatever the progress of our science in the future.

We, who have devoted our lives to the solution of problems connected with physics, now meet together to help each other and to forward the interests of the subject which we love. A subject which appeals most strongly to the better instincts of our nature, and the problems of which tax our minds to the limit of their capacity and suggest the grandest and noblest ideas of which they are capable.

In a country where the doctrine of the equal rights of man has been distorted to mean the equality of man in other respects, we form a small and unique body of men, a new variety of the human race, as one of our greatest scientists calls it, whose views of what constitutes the greatest achievement in life are very different from those around us. In this respect we form an aristocracy, not of wealth, not of pedigree, but of intellect and of ideals, holding him in the highest respect who adds the most to our knowledge or who strives after it as the highest good.

Thus we meet together for mutual sympathy and the interchange of knowledge, and may we do so ever with appreciation of the benefits to ourselves and possibly to our science. Above all, let us cultivate the idea of the dignity of our pursuit, so that this feeling may sustain us in the midst of a world which gives its highest praise, not to the investigation in the pure etherial physics which our society is formed to cultivate, but to the one who uses for satisfying the physical rather than the intellectual needs of mankind. He who makes two blades of grass grow where one grew before is the benefactor of mankind; but he who obscurely worked to find the laws of such growth is the intellectual superior as well as the greater benefactor of the two.

How stands our country, then, in this respect? My answer must still be now as it was fifteen years ago, that much of the intellect of the country is still wasted in the pursuit of so-called practical science which ministers to our physical needs and but little thought and money is given to the grander portion of the subject which appeals to our intellect alone. But your presence here gives evidence that such a condition is not to last forever.

Even in the past we have the names of a few whom scientists throughout the world delight to honor. Franklin, who almost revolutionized the science of electricity by a few simple but profound experiments. Count Rumford, whose experiments almost demonstrated the nature of heat. Henry, who might have done much for the progress of physics had he published more fully the results of his investigations. Mayer, whose simple and ingenious experiments have been a source of pleasure and profit to many. This is the meager list of those whom death allows me to speak of and who have earned mention here by doing something for the progress of our science. And yet the record has been searched for more than a hundred years. How different had I started to record those who have made useful and beneficial inventions!

But I know, when I look in the faces of those before me, where the eager intellect and high purpose sit enthroned on bodies possessing the vigor and strength of youth, that the writer of a hundred years hence can no longer throw such a reproach upon our country. Nor can we blame those who have gone before us. The progress of every science shows us the condition of its growth. Very few persons, if isolated in a semi-civilized land, have either the desire or the opportunity of pursuing the higher branches of science. Even if they should be able to do so, their influence on their science depends upon what they publish and make known to the world. A hermit philosopher we can imagine might make many useful discoveries. Yet, if he keeps them to himself, he can never claim to have benefited the world in any degree. His unpublished results are his private gain, but the world is no better off until he has made them known in language strong enough to call attention to them and to convince the world of their truth. Thus, to encourage the growth of any science, the best thing we can do is to meet together in its interest, to discuss its problems, to criticize each other's work and, best of all, to provide means by which the better portion of it may be made known to the world. Furthermore, let us encourage discrimination in our thoughts and work. Let us recognize the eras when great thoughts have been introduced into our subject and let us honor the great men who introduced and proved them correct. Let us forever reject such foolish ideas as the equality of mankind and carefully give the greater credit to the greater man. So, in choosing the subjects for our investigation, let us, if possible, work upon those subjects which will finally give us an advanced knowledge of some great subject. I am aware that we cannot always do this: our ideas will often flow in side channels: but, with the great problems of the Universe before us, we may sometime be able to do our share toward the greater end. . . .

All the facts which we have considered, the liability to error in whatever direction we go, the infirmity of our minds in their reasoning power, the fallibility of witnesses and experimentors, lead the scientist to be specially skeptical with reference to any statement made to him or any so-called knowledge which may be brought to his attention. The facts and theories of our science are so much more certain that those of history, or of the testimony of ordinary people on which the facts of ordinary history or of legal evidence rest, or of the value of medicines to which we trust when we are ill, indeed to the whole fabric of supposed truth by which an ordinary person guides his belief and the actions of his life, that it may seem ominous and strange if what I have said of the imperfections of the knowledge of physics is correct. How shall we regulate our mind with respect to it: there is only one way that I know of and that is to avoid the discontinuity of the ordinary, indeed the so-called cultivated legal mind. There is no such thing as absolute truth and absolute falsehood. The scientific mind should never recognize the perfect truth or the perfect falsehood of any supposed theory or observation. It should carefully weigh the chances of truth and error and grade each in its proper position along the line joining absolute truth and absolute error.

The ordinary crude mind has only two compartments, one for truth and one for error; indeed the contents of the two compartments are sadly mixed in most cases; the ideal scientific mind, however, has an infinite number. Each theory or law is in its proper compartment indicating the probability of its truth. As a new fact arrives the scientist changes it from one compartment to another so as, if possible, to always keep it in its proper relation to truth and error. Thus the fluid nature of electricity was once in a compartment near the truth. Faraday's and Maxwell's researches have now caused us to move it to a compartment nearly up to that of absolute error. . . .

Yet it would be folly to reason from this that we need not guide our life according to the approach to knowledge that we possess. Nature is inexorable; it punishes the child who unknowingly steps off a steep precipice quite as severely as the grown scientist who steps over, with full knowledge of all the laws of falling bodies and the chances of their being correct. Both fall to the bottom and in their fall obey the gravitational laws of inorganic matter, slightly modified by the muscular contortions of the falling object, but not in any degree changed by the previous belief of the person. Natural laws there probably are, rigid and unchanging ones at that. Understand them and they are beneficent: we can use them for our purposes and make them the slaves of our desires. Misunderstand them and they are monsters who may grind us to powder or crush us in the dust. Nothing is asked of us as to our belief: they act unswervingly and we must understand them or suffer the consequences. Our only course, then, is to act according to the chances of our knowing the right laws. If we act correctly, right; if we act incorrectly, we suffer. If we are ignorant we die. What greater fool, then, than he who states that belief is of no consequence provided it is sincere.

An only child, a beloved wife, lies on a bed of illness. The physician says that the disease is mortal; a minute plant called a microbe has obtained entrance into the body and is growing at the expense of its tissues, forming deadly poisons in the blood or destroying some vital organ. The physician looks on without being able to do anything. Daily he comes and notes the failing strength of his patient and daily the patient goes downward until he rests in his grave. But why has the physician allowed this? Can we doubt that there is a remedy which shall kill the microbe or neutralize its poisons? Why, then, has he not used it? He is employed to cure but has failed. His bill we cheerfully pay because he has done his best and given a chance of cure. The answer is *ignorance*. The remedy is yet unknown. The physician is waiting for others to discover it or perhaps is experimenting in a crude and unscientific manner to find it. Is not the inference correct, then, that the world has been paying the wrong class of men? Would not this ignorance have been dispelled had the proper money been used in the past to dispell it? Such deaths some people consider an act of God. What blasphemy to attribute to God that which is due to our own and our ancestors' selfishness in not founding institutions for medical research in sufficient numbers and with sufficient means to discover the truth. Such deaths are murder. Thus the present generation suffers for the sins of the past and we die because our ancestors dissipated their wealth in armies and navies, in the foolish pomp and circumstance of society, and neglected to provide us with a knowledge of natural laws. In this sense they were the murderers and robbers of future generations of unborn millions, and have made the world a charnel house and a place of mourning where peace and happiness might have been. Only their ignorance of what they were doing can be their excuse, but this excuse puts them in the class of boors and savages who act according to selfish desire and not to reason and to the calls of duty. Let the present generation take warning that this reproach be not cast on it, for it cannot plead ignorance in this respect.

This illustration from the department of medicine I have given because it appeals to all. But all the sciences are linked together and must advance in concert. The human body is a chemical and physical problem, and these sciences must advance before we can conquer disease.

But the true lover of physics needs no such spur to his actions. The cure of disease is a very important object and nothing can be more noble than a life devoted to its cure.

The aims of the physicist, however, are in part purely intellectual: he strives to understand the Universe on account of the intellectual pleasure derived from the pursuit, but he is upheld in it by the knowledge that the study of nature's secrets is the ordained method by which the greatest good and happiness shall finally come to the human race.

Where, then, are the great laboratories of research in this city, in this country, nay in the world? We see a few miserable structures here and there occupied by a few starving professors who are nobly striving to do the best with the feeble means at their disposal. But where in the world is the institute of pure research in any department of science with an income of $100,000,000 per year? Where can the discoverer in pure science earn more than the wages of a day laborer or cook? But $100,000,000 per year is but the price of an army or of a navy designed to kill other people. Just think of it, that one per cent of this sum seems to most people too great to save our children and descendents from misery and death!

But the twentieth century is near, may we not hope for better things before its end? May we not hope to influence the public in this direction? . . .

INDEX